Targeted Therapies in Cancer: Myth or Reality?

ADVANCES IN EXPERIMENTAL MEDICINE AND BIOLOGY

Editorial Board:

NATHAN BACK, *State University of New York at Buffalo*
IRUN R. COHEN, *The Weizmann Institute of Science*
ABEL LAJTHA, *N.S. Kline Institute for Psychiatric Research*
JOHN D. LAMBRIS, *University of Pennsylvania*
RODOLFO PAOLETTI, *University of Milan*

Recent Volumes in this Series

Volume 602
OSTEOIMMUNOLOGY: INTERACTIONS OF THE IMMUNE AND SKELETAL SYSTEMS
  Edited by Yongwon Choi

Volume 603
THE GENUS YERSINIA: FROM GENOMICS TO FUNCTION
  Edited by Robert D. Perry and Jacqueline D. Fetherson

Volume 604
ADVANCES IN MOLECULAR ONCOLOGY
  Edited by Fabrizio d'Adda di Gagagna, Susanna Chiocca, Fraser McBlane and Ugo Cavallaro

Volume 605
INTEGRATION IN RESPIRATORY CONTROL: FROM GENES TO SYSTEMS
  Edited by Marc Poulin and Richard Wilson

Volume 606
BIOACTIVE COMPONENTS OF MILK
  Edited by Zsuzsanna Bosze

Volume 607
EUKARYOTIC MEMBRANES AND CYTOSKELETON: ORIGINS AND EVOLUTION
  Edited by Gáspár Jékely

Volume 608
BREAST CANCER CHEMOSENSITIVITY
  Edited by Dihua Yu and Mien-Chie Hung

Volume 609
HOT TOPICS IN INFECTION AND IMMUNITY IN CHILDREN VI
  Edited by Adam Finn and Andrew J. Pollard

Volume 610
TARGETED THERAPIES IN CANCER
  Edited by Francesco Colotta and Alberto Mantovani

A Continuation Order Plan is available for this series. A continuation order will bring delivery of each new volume immediately upon publication. Volumes are billed only upon actual shipment. For further information please contact the publisher.

Francesco Colotta
Alberto Mantovani

Editors

Targeted Therapies in Cancer:

Myth or Reality?

 Springer

Francesco Colotta
Nerviano Medical Sciences
10 Viale Pasteur
Nerviano (Milano) 20014, Italy
francesco.colotta@nervianoms.com

Alberto Mantovani
Istituto Clinico Humanitas and
 Institute of Pathology
University of Milan
56 Via Manzoni
Rozzano, Milano 20089, Italy
alberto.mantovani@humanitas.it

ISBN 978-0-387-73897-0   e-ISBN 978-0-387-73898-7

Library of Congress Control Number: 2007936362

© 2008 Springer Science+Business Media, LLC
All rights reserved. This work may not be translated or copied in whole or in part without the written permission of the publisher (Springer Science+Business Media, LLC, 233 Spring Street, New York, NY 10013, USA), except for brief excerpts in connection with reviews or scholarly analysis. Use in connection with any form of information storage and retrieval, electronic adaptation, computer software, or by similar or dissimilar methodology now known or hereafter developed is forbidden.
The use in this publication of trade names, trademarks, service marks, and similar terms, even if they are not identified as such, is not to be taken as an expression of opinion as to whether or not they are subject to proprietary rights.

Printed on acid-free paper.

9 8 7 6 5 4 3 2 1

springer.com

Nerviano Medical Sciences (www.nervianoms.com) was intended as a forum for scientists and clinicians working in cancer drug discovery and therapy to share their reflections and experiences on how the paradigm shift from empiricism to molecular-targeted therapies is contributing to the translation of basic knowledge into new therapies for cancer patients. This book collects the contributions given by scientists and clinicians, from academia and industry, who participated in this meeting.

We hope that this book contributes to the improvement of our approach to cancer drug discovery and helps us find new, more efficacious and better tolerated drugs for cancer patients. It provides an overview of diverse approaches ranging from drug discovery to cellular therapy. Although this change in paradigm has been useful, its entry into the clinical arena was associated with unforeseen problems including the emergence of resistance, unexpected side effects and failures. Time is, therefore, ripe for a critical cultural reflection on the state of the art, prospects and limitations. Ultimately, is targeted therapy in cancer a myth or a reality?

Francesco Colotta
Alberto Mantovani

Introduction

Cancers share a restricted set of characteristics crucial to the tumor phenotype: proliferation in the absence of external growth stimuli, avoidance of apoptosis and no limits to replication, escape from external growth-suppressive forces and the immune response, an inflammatory micro-environment with new blood vessel formation, and an ability to invade normal tissues. In the last 20 years, the molecular determinants of these behaviors are becoming increasingly well understood. This has changed the current paradigm underlying the drug discovery process intended to identify novel therapies to fight cancer.

In the past most efforts to identify novel therapies to cancer were focused on the empirical observation of natural or chemically synthesized molecules that inhibited cancer cell growth in vitro and/or in vivo. Most often, the molecular mechanisms underlying the observed anti-cancer activities of empirically discovered anti-cancer agents were discovered afterwards. Cornerstones of currently used chemotherapeutic armamentarium have been discovered according to this "empiricism-based paradigm."

The "molecular target paradigm" is focused on the molecular determinants of aberrant cancer behavior. Conceptually, this paradigm starts with the identification and molecular characterization of proteins that are mutated or over-expressed in cancer cells, and that are believed to play a key role in cancer cell biology. High throughput screening and modern medicinal chemistry, along with sophisticated techniques like computational chemistry and modeling, lead to rapidly identifying hits and then leads that specifically and potently inhibit the activity of proteins mutated or over-expressed in cancer. Recombinant approaches have also been successfully used to generate molecular-targeted biologicals that specifically hit proteins aberrantly expressed in cancer cells.

The global effort sustained by the scientific community and the pharmaceutical industry to discover new approaches to fight cancer is impressive. Thousands of scientists are devoted to this mission and the global investment is in the order of several billion Euros. The "targetcentric" paradigm in cancer drug discovery is widely accepted and used in the pharmaceutical industry and represents the current standard approach to cancer drug discovery.

In September 2005 an International Meeting on "Targeted Therapies in Cancer: Myth or Reality" was held in Milan. This successful meeting sponsored by

Contents

Introduction v

Causality in Medicine 1
Giulio Giorello

The Evolution of the Biomedical Paradigm in Oncology:
Implications for Cancer Therapy 5
Gilberto Corbellini and Chiara Preti

Anticancer Drug Discovery and Development 19
Francesco Colotta

Beyond VEGF: Targeting Tumor Growth and Angiogenesis
via Alternative Mechanisms 43
James Christensen and Kenna Anderes

Aurora Kinases and Their Inhibitors: More
Than One Target and One Drug 54
Patrizia Carpinelli and Jürgen Moll

Signalling Pathways and Adhesion Molecules as Targets
for Antiangiogenesis Therapy in Tumors 74
Gianfranco Bazzoni

Developing T-Cell Therapies for Cancer in an Academic Setting 88
Malcolm K. Brenner

**Anticancer Cell Therapy with TRAIL-Armed CD34⁺
Progenitor Cells** 100
Carmelo Carlo-Stella, Cristiana Lavazza, Antonino Carbone,
and Alessandro M. Gianni

**Linking Inflammation Reactions to Cancer: Novel Targets
for Therapeutic Strategies.** 112
Alberto Mantovani, Federica Marchesi, Chiara Porta,
Paola Allavena, and Antonio Sica

**Clinical Development of Epidermal Growth Factor Receptor (EGFR)
Tyrosine Kinase Inhibitors: What Lessons Have We Learned?** 128
Manuel Hidalgo

**GIST As the Model of Paradigm Shift Towards Targeted Therapy
of Solid Tumors: Update and Perspective on Trial Design** 144
Jaap Verweij, Caroline Seynaeve, and Stefan Sleijfer

**Monoclonal Antibodies in the Treatment of Malignant
Lymphomas** 155
Bertrand Coiffier

**Molecular Network Analysis using Reverse Phase Protein
Microarrays for Patient Tailored Therapy** 177
Runa Speer, Julia Wulfkuhle, Virginia Espina, Robyn Aurajo,
Kirsten H. Edmiston, Lance A. Liotta, and Emanuel F. Petricoin III

Index 187

1
Causality in Medicine

Giulio Giorello

In the words of the great astronomer John Frederick William Herschel [son of Friedrich Wilhelm, the scientist of great renown who, with his sister Caroline Lucretia, discovered the planet Uranus (1781), author of the fundamental text of philosophy of science *Preliminary Discourse on the Study of Natural Philosophy* (1830)], when a "mind" studies any phenomenon that gives rise to a sensation of wonder or even fear, then the next step must be to find a way to discover what produced that sensation. This, of course, is a hallowed tradition: the famous motto of Vergil comes to mind, "felix qui potuit rerum cognoscere causas" (*Georgics*, II, 490), that motto which paraphrases the well-known passage penned by Lucretius in his De rerum natura (III, 1072). This statement, however, must be revised when taken in a typically empirical context: the clearest indication that the "true cause" has been identified is when it not only provides an answer to the original question, but also offers an explanation of many other facts, sometimes exceeding even the wildest hopes of the original researchers. (It may even be that on re-examination what was originally thought to be evidence **against** a certain hypothesis may turn out to be quite the opposite).

Herschel, in the tradition of his times, defined as **natural philosophy** that which today we call more simply **science**; but this term caught the Lucretian (or Vergilian) spirit of the attempt to return "anomalous" events to the regularity of a pattern which often, though not always, justified rational action. If (a) the cause has been identified **and** (b) **it is possible intervene**, then it is perfectly reasonable to expect a (desired) change in the effects. This is the logic on which the Greeks based their *techne* and the Latins their *ars*, and the art of medicine (*ars medica*) can also be made to fit into this framework. The way in which this has been done over time represents one of the most important achievements of scientific and philosophical thought.

It is not necessary to cite all the classic precedents; it is, in fact, sufficient to mention Claude Bernard's well-known *Introduction à l'étude de la médicine expérimentale* (1865) and the excellence of the empiricist approach that he

---

Università degli Studi di Milano, Dipartimento di Filosofia, Via Festa del perdono 7 - 20122, Milano, g.giorello@tiscali.it

delineates in its pages. The aim of this discipline is "to preserve health and cure illness." The latter is, in fact, those very "abnormalities" that provoke wonder (and of which, understandably, people tend to be afraid); it is necessary to know the causes of the biological phenomena in the normal state ("physiology") to explain them. This discipline teaches us to maintain the natural conditions of life as they are, and by so doing, to conserve health; however it is equally important to have a working knowledge of illness and "the factors that provoke illness" ("pathology"): by this means we are able to impede the onset of illness, or at least to fight its effects with various therapies – in short, it allows us to recover.

Bernard does not restrict his considerations to the simple "experience" of the sick and of successful healers. Like all good empiricists, and in no way differing from the great theoreticians like Herschel, for his specific field - which is experimental medicine as a support to therapy - he maintains that the observation of the pertinent phenomena should be integrated with conjectures which, in turn, have to be checked with experiments, with true empirical tests. If the practitioner is to be a true scientist he must first ascertain the facts and formulate a hypothesis; then ideate an experiment and invent the conditions by which to test his hypothesis. This experiment will give birth to new phenomena that will draw his attention and cause him to formulate further hypotheses, and so on. Therefore, we can see that medical knowledge is an "active" science rather than a mere registration of facts, and this long before it is applied to each individual patient ("therapy"). This is why disciplines such as "physiology" and "pathology" are so important. Bernard flanks his observations of the conditions of each individual patient with his laboratory work. This "active" aspect of "experimental medicine" allows the practitioner to adjust the errors in every day medical practice. He himself wrote that: "Science […]rejects **that which is not determined**, and when, as in medicine, one wishes to base one's opinions on one's clinical eye, on inspiration or a somewhat vague intuition, then one is outside the bounds of science and provides an example of bizarre medicine which can be the cause of the very worst forms of damage as it entrusts the health and life of the patient to the vagaries of ignorant visionaries. True science teaches us to doubt and not to take decisions when knowledge is not present." In his work entitled *Biologia e medicina tra molecole, formazione e storia* (Laterza, Roma-Bari 1991), Giovanni Felice Azzone indicates that it was Claude Bernard who proposed the causal paradigm in medicine, in a sophisticated process of model building, in which the models were to be controlled by experiments. This is, in fact, the very aspect which changes the nature of the intervention in the context of prevention and in the context of therapy. For Claude Bernard this type of intervention is "rational" insofar as it is justified by a type of "determinism" that allows us to predict the effects given the cause. It must be kept in mind, however, that this form of determinism must be interpreted as a methodological premise and not as a metaphysical foundation: without this premise, science would be completely "impotent" – and an unbridgeable abyss would open up between medical theory and practice, just as it was in the days of "empirical medicine,",which is very different from "experimental medicine."

1 Causality in Medicine

The causal paradigm has been seen to be extremely flexible. Claude Bernard's framework permits fruitful analysis of situations in which more than one "next cause" must be dealt with. There is, for example, the interesting historical case of malaria. Following the example of Bernardino Fantini, medical historian, in his excellent reconstruction of the case, we will limit our account by recalling that at the end of the 19th Century it had been established that a parasite of the genus *plasmodium* was the specific and necessary cause of malaria, while Ronald Ross and Battista Grassi had indicated that the vector of this disease was an *anopheles* mosquito. Traditionally, the so-called "Camillo Golgi's law" ("man + plasmodium = malaria") is considered to be in contrast with "Battista Grassi's law" ("man + anopheles = malaria"). Actually, these are two points of view at two different levels: "Golgi's law" deals with the clinical aspects of malaria, while "Grassi's law" regards the epidemiology of the disease. As is well known, malaria can develop in the absence of anopheles (when inoculated, for example), but it will not develop in the absence of plasmodium. The presence of anopheles is a necessary condition for the disease to spread endemically or epidemically in any given region. This is why travelers who have contracted malaria in areas in which the disease is habitually rife do not constitute a risk when they return to countries in which anopheles are not present. (See B. Fantini, "Anophelism without malaria: an ecological and epidemiological puzzle", in *Parassitologia*, 36, 1994, pp. 83–106).

Given the widespread incidence and the degenerative aspects of the pathologies which are generally known as **cancer** (Rita Levi-Montalcini once defined it as a form of Milton-like rebellion of the body's cells), it is obvious that the causal paradigm played an increasingly relevant role both at the theoretical research level and in prevention and therapy. A classic case in the literature comes to mind. There is a Mayan sculpture in stone in the National Museum del Guatemala, which represents a victim of the so-called retinoblastoma; this dramatic sculpture clearly shows the devastating effects on the eye of a retinal cancer in the advanced stages. In Europe the theory of this disease was studied as early as the 16th Century. Over time various causes have been cited for it ranging from fungal infections and trauma to over-exposure to candle light! "Knowing" the causes, however, did not automatically enable practitioners to intervene on the effects at a practical level. The aspects (a) and (b) of the causal framework were dramatically separate. Some practitioners sustained that the eye should be removed, but how could such a terrible operation be carried out at a time when anesthetics were unknown? This was an instance in which the medicine that Bernard later considered as experimental science was still "impotent;" more so because children with retinoblastoma did not survive their early years. It was only after the ophthalmoscope and the use of ether were discovered that practitioners were able to diagnose this tumor in the eye when it was still in the very early stages and operate to save their patients. As a consequence, some retinoblastoma children did survive to maturity, though partially or completely blind, and had offspring; this made it possible to discover that the same form of cancer of the eye also developed in their children. However, this was only a "regularity" that gave rise to concern, not a "deterministic" law in the strict sense of the term as it was understood by Claude Bernard. In certain cases of retinoblastoma, there was no heredity involved while the offspring of some of the survivors of the disease were perfectly healthy from this point of view. Moreover there were a significant number of retinoblastoma victims whose parents did not suffer from the disease, but whose children did. At the close of the 1960s a number of audacious hypotheses were formulated regarding this mystery of retinoblastoma; one of the most outstanding was that proposed by Alfred Knudson, who was studying at the University of Texas at the time.

> Retinoblastoma, he said, was a two-hit disease. A single retinal cell had to be battered by two mutations to its DNA before it would turn cancerous. Knudson proposed that those children who had multiple tumors early in life were born with a built-in defect – their first "hit" was part of their genetic legacy. They carried the retinoblastoma trait and, thus, their eye cells had to undergo only one more mutational hit before cancer developed. The children who fell into the category of retinoblastoma carriers were of two types: either they inherited the trait from a retinoblastoma parent, or the chromosomes of the mother or father had suffered mutational damage during the development of the egg or sperm cell. In either instance, the offspring were bestowed with the retinoblastoma trait at the moment of conception and, thus, were extremely susceptible to the disease. By contrast, the older children who had only one tumor suffered from so-called sporadic retinoblastoma. They began with perfectly normal chromosomes, but at some point during the growth of the eye a lone retinal cell had the great misfortune to be struck twice. Two times in the course of ocular development, the DNA of one cell was wounded – perhaps by a mutagen or cosmic ray, perhaps by the random errors in chromosomal replication. The doubly damaged cell then proliferated into a tumor.

This is an extract from the work *Natural Obsessions* that Natalie Angier dedicated to the search for the oncogene (Houghton Mifflin Company, Boston 1988, pp. 329–330). Alfred Knudson's explanation clearly illustrates the sense of a **causal model** that embeds "physiology" and "pathology" at a theoretical level in Claude Bernard's sense of the framework supplied by the so-called **Neodarwinism** – i.e., the Darwinian synthesis of evolution, genetics and molecular biology. As the great biologist Ernst Mayr rightly observed, every time we have recourse to any level of an evolutionistic type of explanation, we are obliged to take into account a certain quota of randomness. Claude Bernard, that advocate of the concept of strict determinism, did not like having recourse to statistics, but today this discipline is rigorously included in the context of the theory of probabilities. Therefore, the random aspects of the mutations we have to consider when individuating the various oncogenes are not an insurmountable conceptual difficulty for the causal paradigm in medicine. The hope is that causal and evolutionistic paradigms may be used together – and this, I believe, is one of the themes of the **present volume**.

2
The Evolution of the Biomedical Paradigm in Oncology: Implications for Cancer Therapy

Gilberto Corbellini and Chiara Preti

2.1 Introduction

According to the view of Harold Varmus, it is time for "changes in the culture of oncology" (Varmus 2006, 1165). The former NIH director and Nobel prize recipient for physiology and medicine for the discovery of oncogenes believes that "during most of the past 50 years, pharmaceutical chemistry continued to serve cancer patients much more effectively than did cancer biology." He argues that, as a consequence of the strategy adopted, "laboratory-based investigations into the nature of cancer cells and clinical efforts to control cancer often seemed to inhabit separate worlds" (Varmus, 2006, 1162). So he points out that "the new era in cancer research" needs "stronger working relationships between bench scientists and their clinical colleagues, between oncologists in academia and those in community hospitals, and between oncologists and other physicians." Moreover "new training programs" should "provide graduate students in the basic sciences with an opportunity to understand the dilemmas posed by cancer as a human disease" (Varmus 2006, 1165).

Varmus' analysis joins the increasing attacks on cancer research policy inspired by the U.S. President Richard Nixon's National Cancer Act of December 1971, declaring the "war on cancer". In a recent issue of *Fortune* Clifton Leaft, who personally experienced the condition of cancer patient, published a long article, "Why we are losing the war on cancer [and how to win it]," reporting statistical data and interviews with leading scientists. Among them, a very negative judgment about the methodological foundation of cancer research came from Robert Weinberg, who told the columnist that experimental oncologists cultivate the "illusion" of doing "something meaningful" just because they can manage straightforward experiments to accumulate a huge amount of reproducible data (Leaft, 2004, 85).

The history of the "war on cancer" has been anatomized and declared a "failure" by an outstanding clinician and oncologist, Guy B. Faguet, who was in the forefront and now is retired. He demonstrates, in his very well received and reviewed book that "the three crucial measures of progress in the War on cancer, cure rates, prolongation of survival, and quality of life, remain stagnant despite enactment of the National

---

Section of History of Medicine, Sapienza – University of Rome, giberto.corbellini@uniroma1.it

F. Colotta and A. Mantovani (eds.), *Targeted Therapies in Cancer*.
© Springer 2008

Cancer Act of 1971" (Faguet 2005, 52). He explains the failures as caused by "an unbalanced focus on treatment of operable cancer to the detriment of prevention and early detection, and adherence to the infectious disease model that has driven drug development towards the cancer cell-kill paradigm" (Faguet 2005).

In this paper we show that the existing main streams of cancer research and treatment reflect the schizophrenic epistemological status of scientific medicine. The field is prompted by two different, dissonant and incomplete philosophies – the bioexperimental and the clinical-epidemiological approaches – for the understanding and management of disease. We demonstrate that the "cancer cell-kill paradigm" that, according to Faguet, misrepresented the cancer problem embodies the epistemological essentialism of the bioexperimental tradition of medicine, which is no more maintainable. At the same time, the clinical-epidemiological approach is in some way threatening the scientific foundation of medical reasoning, as it spread among students the idea that statistical correlations can replace causal explanation (Thagard 1999). Then we argue that a new theoretical perspective is emerging in cancer biology, the evolutionary or Darwinian model of cancer progression, which indicates a more dynamic and realistic view of cancer, and highlights new paths of discovery for cancer therapy. Our original contribution, as historian and philosopher, is a reconstruction of the main conceptual steps that led to the understanding of cancerogenesis as a Darwinian process.

2.2 The Epistemological Evolution of Scientific Medicine

Let us introduce a very schematic view of the epistemological evolution of medicine. The main historical traditions of Western medicine are the clinical, the physiopathological or bioexperimental and the clinical-epidemiological (Corbellini, 2007). The clinical paradigm emerged with Hippocratic medicine and lasted until the beginning of the 20th Century. According to early and modern clinicians the knowledge of disease can be attained by observing and interpreting a patient's natural or artificially induced symptoms and signs. From the late 17th Century a medical discipline called nosology emerged to classify symptoms and signs and to create specific patterns of disease entities, useful for making diagnoses (Porter 1997).

During the second half of the 19th Century, the founders of scientific medicine were able to exploit the new physiological and microbiological knowledge by means of the systematic application of the experimental method. They created the physiopathological, or also so-called bioexperimental paradigm, based on the idea that the knowledge of disease must aim at developing explanatory theories and experimental models to identify the proximate causes producing functional alterations or disease. Physiopathologists assumed that heterogeneity and individual variations of data, that ancient clinicians explained assuming an individual constitution or diathesis for each patient, depended on some intrinsic limitations of the experimental models, and were a sort of noise. According to the followers of Claude Bernard (1865), disease is a deviation from functional homeostasis that can be caused by several factors – internal and external to the organism. They defined health as absence of disease, establishing as a goal of medicine the treatment or prevention of diseases by rationally designed drugs and interventions, based on the understanding of the etiopathogenesis.

The bio-experimental paradigm has been at the origins of the greatest results of medicine, in terms of basic knowledge and of prevention and control of infectious, hereditary (monogenic) and chronic or degenerative diseases. However, the emerging complexity of the molecular, biochemical and cellular dynamics involved in etiophysiopathology and the increasing frequency of chronic-degenerative diseases, with their multiple and statistically defined determinants, created a less favorable environment for the biomedical model, that suffered a decline of effectiveness (Corbellini 2007).

A new paradigm emerged, thanks to the improvement of statistical analysis of experimental design. The new perspective came out with the invention of the clinical trials, in the second half of the 1940s, that stimulate the rise of clinical epidemiology, and later the advent of the evidence-based medicine movement (Corbellini 2007). The implementation and success of clinical trials in assessing the effectiveness of drugs brought to the view that there is no need to know the functional mechanisms that cause the clinical phenomena: clinicians can just apply statistical methods to inductively establish causal correlations. Clinical trials based on frequentist statistics, has become the only reliable experimental design to test hypotheses and evaluate the effectiveness and efficiency of medical decisions.

This is not the seat to analyze the epistemological controversies between the bioexperimental and the clinical-epidemiological paradigms that live together in today's biomedical world, which means in medical faculties and literature (Corbellini 2007). They are both practically very useful. However, they are both based on an incomplete view of biomedical phenomena and seem not able to see their intrinsic epistemological limits, and they lack an historical or evolutionary perspective of diseases and health. In fact, they ignore the implications of the biological fact that, due to phylogenetic and ontogenetic causes, the phenotypic traits of an organism, including heath and disease, result from individual histories (Corbellini 2007).

In the light of the previous picture we can now better understand how cancer therapy evolved.

2.3   From Magic Bullets to Targeted Therapies: Many Treatments but the Same Philosophy

The origins of modern cancer therapy can be seen as one of the main instantiations of the biomedical paradigm. In fact, Paul Ehrlich's view (1906-9) of specific chemotherapy as a *therapia sterilisans magna* and of drugs as artificially designed antibodies or "magic bullets" has inspired the search for effective anticancer drugs since the beginnings. Cancer chemotherapy started in 1946 when Goodmann and

Gilman observed a dramatic reduction in tumor mass of a patient with non-Hodgkin lymphoma after the injection of a chemical derivate of nitrogen mustard (Goodmann and Gillman, 1946). Cancer chemotherapy has gone through several different phases, schematically listed in Fig. 2-1, characterized by important technical novelties, but not by any cultural change. According to Varmus (2006, 1162) "targeted therapy, in a sense, are not more targeted than the conventional chemotherapies," while the classical assumptions of chemotherapy and drug discovery has been challenged by several new findings and concepts.

Fig. 2-1 An essential timeline of cancer therapy

1946–1950	Serendipitous discovery of anticancer activity of nitrogen mustard derivates and of synthetic antifolates (MTX): cancer chemotherapy can be pursued.

1950–1960	Methodological foundation of cancer chemotherapy, development of in vitro and in vivo model, and discovery of new anticancer drugs mainly by empirically testing natural and artificial products
	    6-mercaptopurine (6-MP)
	    vincrastine
	    cyclophosphamide

1960–1970	The first successes (Hodgkin's lymphoma and ALL) resulted from clinical experiences, supported by new ideas on chemotherapy kinetics
	    MOMP
	    Taxanes
	    Cisplatin
	    adriamycin
	    "Remission induction therapy" or "total therapy" of ALL

1970 – 1980	Adjuvant chemotherapy shows efficacy, endocrine treatment comes of age, but cancer cells hold genetic mechanism to acquire resistance to anticancer drugs
	    ABVD
	    Tamoxifen
	    Resistence of cancer cells to MTX

1980–1990	Setting molecular biotechnology to fighting cancer: genes for target proteins are mapped and cloned, monoclonal antibodies and recombinant vaccines are developed (immunology enters the game)
	    Interferon therapy
	    Recombinant hepatitis B vaccines

1990 –	The rise of targeted therapies (azacytidine, trastuzumab, imitinib mesylate, bevacizumab and gefitinib) and the bright hopes of tailored/personalized therapies
	    Colony stimulating factors
	    Interleukin-2
	    Azacytidine
	    Rituximab
	    Ontak® - recombinant Dna-derived cytotoxic protein
	    Trastuzumab - Herceptin®
	    Imitinib mesylate - Gleevec®
	    Bevacizumab (Avastin®)
	    Gefitinib (Iressa®)
	    Oncotype DX®
	    Gardasil®

The invention/discovery of magic bullets – with exclusive and absolute specificity – has constantly oriented the search for anticancer drugs from the beginning to targeted therapies. However, apart from monoclonal antibodies, the ideal of designing drugs *de novo* or synthesizing tailor-made chemicals that fulfill all requirements of efficacy, tolerance and absorption, distribution, metabolism and excretion characteristics has remained elusive (Drews, 2006). Moreover, the concept of absolute specificity doesn't make much sense from a biological viewpoint – physiological redundancies (degeneracy) and pleiotropy in cell signalling pathways have represented successful evolutionary strategies to develop the most adaptive traits (Searls 2003).

According to most oncologists the eradication of cancer cells, considered like a parasite extraneous to normal physiology of the body, as the final aim of cancer therapy inspired medical treatment of cancer for most of the past century (Faguet 2005). Such a view is no more tenable in the light of cancer molecular genetics, and a more pertinent philosophy should emerge (Reddy and Kaelin, 2002; Faguet, 2005). Paradoxically this view, that embodies the bioexperimental paradigm of medical research, can be maintained thanks to the prevailing influence of a clinical-epidemiological paradigm. In fact, the pivotal role played by the clinical trial to discriminate the levels of therapeutic efficacy of a new treatment scotomize the problem concerning the biological plausibility of drug pharmacological activity.

The new philosophy has to also take into account the expectations that inspire the drug designer, that the serendipitous discovery of new treatments was a temporary consequence of the lack of better knowledge and techniques, and that it is becoming possible to design an effective drug starting from the knowledge of the chemical property of the target, can be misleading (Horrobin 2003). Some experts think that because of the biological nature of the therapeutic targets, drug discovery will always rely on intuition, serendipity and luck, alongside rigorous science and rational thinking (Drews 2006).

2.4 The Darwinian paradigm in oncology

In our opinion, oncology is going through a theoretical revolution that challenges the dominant paradigms. Oncology – as with other branches of biomedicine like, in the past, immunology, medical microbiology and some aspects of neurobiology – is acquiring a more coherent biological way of thinking about the causal dynamics of adaptive physiological and pathological phenomena. This change may still support new therapeutic strategies, but certainly allows better integrating of basic and applied cancer research. It is foreseeable that in the near future the new paradigm will influence cancer therapy as well.

Let us introduce this idea by quoting from two leading oncologists. According to Robert Weinberg and Douglas Hanahan, the six hallmarks of cancer (self-sufficiency in growth signals, insensitivity to anti-growth signals, evading apoptosis, limitless replicative potentials, sustained angiogenesis, tissue invasion and metastatis) are nothing but "acquired capabilities," and "tumor development proceeds via a process formally analogous to Darwinian evolution, in which a succession of genetic changes, each conferring one or another type of growth advantage, leads to the progressive conversion of normal human cells into cancer cells" (Hanahan and Weinberg, 2000).

The heuristic role played by the evolutionary view of cancer progression has been testified by Bert Vogelstein's fundamental contributions to the understanding of the genetic basis of tumor progression in colorectal cancer, that adopted the hypothesis that tumorigenesis is an evolutionary process (Fearon and Vogelstein 1990). Since then theoretical oncology deals with the concept of "cancer cell evolution." Martin Nowak, who works out mathematical approaches to evolutionary dynamics, has transformed Vogelstein's evolutionary view in a series of equations (Michor, Iwsa and Nowak 2004). Even epidemiologists are debating the usefulness of Darwinism as a theoretical framework to make sense of the role of environmental factors in carcinogenesis (Vines 2006).

So, the most advanced oncological research has reached the agreement that cancers are conventional Darwinian processes of repeated cycles of mutations and selections, and that Darwinian models of cancer progression can explain most tumor phenomenology. Environment contributes to cancer development with mutagenic chemicals and conditions that increase cell replications, thus creating the opportunity for mutations to occur and for the somatic selection of the advantageous ones to take place. Mutations that allow cancer cells to produce their own signals to stimulate mitosis, to suppress contact inhibition, to evade apoptosis, to attract the vascular system and to spread or metastasize can have a selective advantage. Somatic mutation and selection are very important for cancer treatment since that, as tumors, can evolve resistance to chemotherapy (Michor, Nowak, and Iwasa 2006).

Weinberg dedicated several pages to the Darwinian view of cancer in his recent and excellent textbook (Weinberg 2006). At the end of his analysis, Weinberg concluded that "the outlines of the model are undoubtedly true, but its details are very difficult to validate" because of the complexities involved in the multi-step tumorigenesis (Weinberg 2006, 423-4).

Well, but in which sense the Darwinian paradigm in oncology represents the natural outfall of the empirical and theoretical investigation on cancer biology? Which are the origins of the Darwinian view of cancer development? How basic research and clinical observations contributed to demonstrate its heuristic value? Which implications for the strategies of cancer therapies have predicted the founders of this new paradigm?

2.5 The Origins of Oncological Darwinism

Let us think to Darwin's *Origins of Species* (1859). Which is the idea that he put forward and emphasized first? Darwin spent the first two chapters illustrating the reality of individual biological variations, under domestication and in natural species. The discovery of a spontaneously occurring heterogeneity in somatic physiological systems with some kind of ability to change adaptively has also been the first step that led to the idea that some selective mechanisms could operate also to produce adaptive physiological responses to unexpected stimuli. The histories of immunology and neurobiology are the best examples of the successful heuristic role played by Darwinian thinking to explain the physiological dynamics that results in adaptive changes to memorize and learn through experience (Corbellini 2007).

In the history of cancer pathology, too, it was the recognition of the diversity of many properties of cancer cells that led to the view that the only unique property of cancer cells is their expression of multiple variables, and that cancer cells heterogeneity, is the biological prerequisite for tumor progression.

In the early and mid-19th Century, it was well recognized that at the macroscopic level, solid cancers had a heterogeneous appearance. Moreover, during the second half of the 19th Century, Virchow and most pathologists reached the view that any cancer cell is like a monad, invested with the potential to develop in any number of ways (Moss 2003). The first successful experiments to induce carcinogenesis by chemical stimulations of normal tissues, reported in 1915, were aimed at confirming Virchow's hypothesis that chronic irritation was the cause of cancer (Moss 2003).

The cytological hypothesis of somatic mutations proposed by Theodor Boveri in 1914, based on previous observations of aberrant mitotic figures in carcinoma samples made by David von Hansemann in 1890, introduced a new interpretation of the process of experimentally induced carcinogenesis (Boveri 1914). Instead of being the product of the contingent interaction between cells and other cells and between cells and extracellular factors, cancer could rather be determined from within a cell.

In the 1940s several studies led to the idea that cancerogenesis was a "multistage process," involving two or more somatic mutations. Peyton Rous spoke of a "two stage process" involving the action of "provocative" and "actuaring" carcinogens (Rous 1943, 581). Isaac Berenblum and Philippe Shubik, in a series of articles, suggested the existence of at least two different aspects of carcinogenesis: initiation and promotion. An event starts the carcinogenic process; another one accelerates it (1947, 1949). Finally, Peter Armitage and Richard Doll, looking for an explanation about the fact that cancer is mainly a disease of elderly, postulated the existence of six to seven stages for carcinogenesis (Armitage, Doll 1954).

The concept of somatic mutation was temporary obscured by the rise of cancer virology. Stimulated by the ideas of Amédée Borrel, who in 1907 discussed the viral origin cancer, and by the discovery of Peyton Rous, who in 1911 discovered the first cancer virus (Rous Sarcoma Virus, RSV) experimentally inducing tumors in chickens, tumor virologists for decades explored the hypothesis that viruses were the source of the stimuli which transform normal cells into tumor cells. According to Rous (1959, 578), "the somatic mutation hypothesis, after more than half a century, remains an analogy." The same conclusion was announced by other leading oncologists (Burdette 1955; Foulds 1969) until the "oncogene hypothesis" was put forward by Huebner and Todaro in 1969 (Huebner and Todaro 1969). The view that cancer transformation was determined by the expression of viral genes (oncogenes) entered the genome of animals allowed to rescue the concept of somatic mutation. In 1971 Howard Temin reintroduced somatic mutation as a failure of normal cellular differentiation mediated by "protoviruses," that become cancer causing viruses (Temin 1971). As we know, in 1976, out of the blue came the discovery that oncogenes are functional components of the animal genomes and the genetic explanation of cancer became definitively established (Stehelin, Varmus and Bishop 1976). Since then, molecular oncology became the leading area of cancer research, bringing to light the proximate causes that concur to produce the complex phenomenology of cancer progression.

The year 1971 was a crucial date for the advancement of cancer genetics. The geneticist Alfred Knudson wrote his landmark paper suggesting the "two-hit" model (completed by David Comings in the 1973 and by Knudson himself in 1985) in order to explain the puzzle of familial and sporadic forms of the retinoblastoma tumor (Knudson 1971; 1985; Cummings 1973). According to Knudson, "the origin of cancer by a process that involves more than one discreet stage is supported by experimental, clinical, and epidemiological disease" (p. 820). He pointed out that, usually, the mutation's number varies "from 3 – 7" (p. 820), and that he wanted to support the hypothesis that "at least one cancer (the retinoblastoma observed in children) is caused by two mutational events" (p. 820). Knudson's analysis of the age-specific incidence of retinoblastoma led him to propose that two "hits" or mutagenic events were necessary for retinoblastoma development. Retinoblastoma occurs sporadically in most cases but, in some families, it displays an autosomal dominant inheritance. In an individual with the inherited form of the disease, Knudson proposed that the first hit is present in the germ line, and thus in all cells of the body. However, the presence of a mutation at the susceptibility locus was argued to be insufficient for tumor formation. Given the high likelihood of a somatic mutation occurring in at least one retinal cell during development, the dominant inheritance pattern of retinoblastoma in some families could be explained. In the non-hereditary form of retinoblastoma, both mutations were proposed to arise somatically within the same cell. Although each of the two hits could theoretically have been in different genes, subsequent studies led to the conclusion that both hits were at the same locus, ultimately inactivating both alleles of the retinoblastoma (RB1) susceptibility gene.

Knudson's hypothesis served not only for illustrating the mechanism through which inherited and somatic genetic changes might collaborate in cancerogenesis, but also to link the concept of recessive genetic determinants for human cancer to somatic cell genetic findings showing the recessive nature of tumorigenesis. In 1969 Harry Harris and his colleagues (Harris, Miller, Klein, Worst and Tachibana 1969), trying to understand whether tumorigenicity was a dominant or recessive trait, proposed that at least one of the chromosomes commonly lost in cases of cancer was a "suppressor" for tumorigenicity. Knudson, in 1983, points out for the first time that cancer might result, not only by activation of oncogenes, but also "through loss of appropriate anti-oncongenes." In this way he was able to justify the retinoblastoma's epidemiological data. Two years later Knudson wrote that "the hereditary cancers have revealed a new class of gene that is important in the pathogenesis of cancer.

The genes of this class, clearly different from oncogenes, have been called antioncogenes, because they produce cancer in a recessive mode, one normal allele being adequate to protect against a particular cancer" (Knudson 1985; 1438).

Physical evidence for the existence of antioncogenes came from the study of several types of tumors, including Wilms tumor, hepatoblastoma, uveal melanoma and bladder cell carcinoma (Koufos, Hansen, Copeland, Jenkins, Lampkin and Cavenee 1986). The RB1 gene was found in 1986 by Friend and his colleagues, which maps to human chromosome 13q14 (Friend, Bernards, Rogelj, et al. 1986).

In the meanwhile that cancer genetics came of age, something was changing in cancer pathology. In 1960, as a consequence of the discovery of cancer stem cells (Wicha, Suling, and Dontu 2006), a few pathologists and clinicians started to disagree with the reductive view of cancer as a disease of the cells. In early 1960s, the British radiologist and radiotherapist Sir David Smithers wrote a long paper in *Lancet* entitled *An attack on cytologism*. He appealed to Karl Popper's falsificationism and criticized the idea that cancer is a defect of cells, suggesting that "cancer is no more a disease of cells than traffic jam is a disease of cars. Cancer is a disease of organization, not a disease of cells" (Smithers 1962; 495).

According to Smithers there is no such a thing as a cancer cell, but only cells behaving in a manner arbitrarily defined as being cancerous. Of course, he criticized somatic mutation hypothesis, as it focused on the idea of oncogenic mutation as an all-or-nothing event, while carcinogenesis is a gradual process. Smithers advanced the idea that an abnormal cell, particularly a stem cell, may produce a clone of cells reacting abnormally with the environment and so promoting the disorganization of the tissues (Smithers 1962).

During the first half of the 20th Century morphological, histological, cytological and physiological investigations on cancer showed that heterogeneity of cancer cells was a common and prominent feature of most human tumors. The number of parameters that could be used to describe cancer cell heterogeneity increased: morphology and function, biochemical markers (then differential expression of gene products), differential growth in vitro, tumorigenicity in vivo (including the number of injected cells required to generate tumors in animals), latency period, growth rate, antigenicity and ability to be affected by a host immune system, ability to invade and metastasize, sensitivity to chemotherapic agents and radiosensitivity (Weiss 2000).

The British pathologist Leslie Foulds was the first to interpret the heterogeneity of cancer cell populations in a dynamic perspective, and was instrumental in explaining tumor progression. In 1949, Foulds defined tumor progression as the "development of a tumor by way of permanent irreversible change in one or more characters of its cells" (Foulds 1949; 373). Actually, he suggested that progression brings the acquisition of new phenotypic traits. In 1958, Foulds described cancer as a "dynamic process advancing through stages that are qualitatively different," progressing from precancerous stages to increasingly invasive and metastasic stages (Foulds 1958; 6). Finally, in a monograph of 1969, Foulds wrote that progression "is not the mere extension of a lesion in space and time but a revolutionary change in a portion of the old lesions establishing a new tumor having properties not formerly manifested" (Foulds 1969; 73).

Even though the increasing evidence of a monoclonal origin of cancer cells seemed to be in contradiction with the apparent heterogeneity, the time was right to reorganize the empirical data using an evolutionary model. Peter Nowell, who had discovered the Philadelphia chromosome, suggested the first explicit Darwinian model of tumor progression in 1976.

In his most quoted paper Nowell (1976) resumed the biological characteristics of tumor progression: a) acquisition by the neoplastic cells of the capacity to invade and to metastasize (malignancy); b) a tendency for neoplastic populations to increase their proliferative capacity (escape from normal growth control mechanisms); c) acquisition of morphologic and metabolic alteration generally interpreted as loss of differentiation; d) maximization of the efficiency in proliferation and invasive growth, and e) elaboration of products which aid the progression (tumor angiogenesis factor).

Finally Nowell suggested that these biological properties represent both the effect of acquired genetic instability in the neoplastic cells, and the sequential selection of variant subpopulations produced as a result of the genetic instability.

2.6 From Speculations to Reality and Beyond: Some Implications of a Darwinian View of Cancer Progression

The confirmation of Nowell's model came from the discovery that the acquisition of the resistance of cancer cells to chemotherapy was similar to other well-known evolutionary phenomena described in medical therapy. In 1978, Robert Schimke discovered a genetic mechanism that provides the condition for a selection of cancer cells resistant to methotrexate (MTX). He showed that resistance of mouse cells to MTXresults from a selection of cells of higher contents of a specific enzyme, due to an increase in the number of copies of the gene coding for this enzyme, that to gene amplification. "The properties of the resistance of cultured cells to MTX, including (i) a stepwise selection of progressively resistant cells; (ii) an increase in a specific protein present at low levels in sensitive cells, which, when present in larger amounts, results in resistance; and (iii) stable and unstable resistance in the absence of selection pressure, have analogies both in antibiotic and insecticide resistance" (Schimke 1978; 1055).

During the 1980s the concepts of genetic instability and clonal evolution were confirmed, and the possibility that epigenetic mechanisms could exert differential selection pressures on heterogeneous cancer cell populations widely discussed. In 1986 Barry Wolman suggested that genetic and chromosomal instability was the potential source of genetic heterogeneity in all tumors, and that variation in local environmental selective pressures and differential survival may contribute to cellular heterogeneity within an expanding tumor. In turn, heterogeneity itself might permit selection and increase in number of aberrant cells which are responsible for tumor progression and metastasis. Genic and chromosomal instability are potential sources for genetic diversity within all tumors. However, variations in local selective forces and differential survival within an expanding solid lesion may contribute to maintenance of a mixed cell population within the primary tumor (Wolman 1986).

Finally, Fearon and Vogelstein (1990) proposed the now historic model of successive genetic changes leading to genetic instability producing colorectal cancer (CRC), in which a number of genes are involved, including APC, k-Ras, DCC, and p53. With few modifications, the Vogelstein model still stands and knowledge on the function and interactions of the key molecules involved, which has been obtained since it was proposed more than 15 years ago, even strengthens the genetic cascade of events in the sequence originally proposed (Weinberg 2006).

According to Vogelstein's group "the genetic instability hypothesis can be viewed as a pessimistic one," as cancer cell heterogeneity should allow the tumors to face therapeutic challenges. However, they think that instability itself could be the Achille's heel, providing a target for drugs killing unstable cells better than normal as it has been demonstrated in the case of yeast cells (Cahill, Kinzler, Vogelstein and Lengauer 1999).

Nowell was the first to put forward that the evolutionary model of cancer progression might induce a pessimistic view about the prospective of a definitive therapeutic success. If the cells within a tumor are so heterogeneous and ready to form variants in the face of therapeutic challenges, do we have a realistic chance of ever curing advanced cancer? According to Nowell (1976) "the fact that most human malignancies are aneuploid and individual in the cytogenetic alterations is somewhat discouraging with respect to the therapeutic considerations" (p. 27). Such a fact explained the failure to discover metabolic alterations sufficient to allow specific chemotherapy. "The same capacity for variation and selection, which permitted the evolution of a malignant population from the original aberrant cell, also provides the opportunity for the tumor to adapt successfully to the inimical environment of therapy, to the detriment of the patient" (p. 27).

Nowell pointed out a further consequence of the clonal evolution model of cancer, that is that each advanced malignancy has individual therapeutic problems, a view that has become adopted by the strategies aimed at developing tailored/personalized therapies (Hasegawa, Ando, Ando, Hashimoto, Imaizumi and Shimokata 2006).

A further implication of a Darwinian model has to do with the fact that any adaptive evolution is, by definition, context dependent. Anderson (2001) thinks that genomic instability should suggest that "instead of directly attacking the heterogeneous population of genomically unstable tumor cells, the invariant, genomically stable cells of the tumor vasculature become an especially appealing target." The idea role of environment was emphasized by Folkman, who saw that solid tumors are angiogenesis-dependent. That brought to the idea of fighting cancer by subtracting blood supply. These ideas have been developed in several lines, one of which brought to the invention of bevacizumab (Ferrara, Hillant, Gerber and Novotny 2004).

Nowell (1976) interpreted the concept of context dependence of cancer in terms of "potential reversibility of the neoplastic process." If the genetically unstable, highly individual malignancy is difficult to eradicate therapeutically, what is the likelihood of producing a "cure" by providing an environment which forces the tumo-r cell population to cease unlimited proliferation and move into a state of controlled differentiation?

Nowell quoted the experiment reported in 1975 by Beatrice Mintz and Karl Illmensee, who injected teratocarcinoma cells taken from embryoid bodies in vivo into developing mouse blastocysts, and obtained normal mice with no evidence of tumors. They, however, found that tumor-derived cells were present in large numbers and contributed to several unrelated tissues. Mintz and Illmensee (1975) concluded that tumor cells were developmentally totipotent and could revert to normal behavior in the appropriate environment. In 1997 Mina Bissel has shown that blocking integrin function was sufficient to revert the malignant phenotype of human breast cancer both in culture and in vivo (Bissel 1997).

More recently, Rudolph Jaenisch and his group demonstrated by using nuclear transplantation that an oocyte's microenvironment can re-establish development pluripotency of malignant cancer cells. The nuclei of murine leukemia, lymphoma and breast cancer cells can support normal preimplantation development to the blastocyst stage, but fail to produce embryonic stem cells. A blastocyst cloned from a RAS-inducible melanoma nucleus develops into ES cells with the potential to differentiate into multiple types in vivo. These findings are in some way paradigmatic for studying the tumorigenic effect of a given cancer genome in the context of the whole animal, and demonstrate that "the malignant phenotype of at least some cancer cells can be reversed to a pluripotent state despite the presence of irreversible genetic alterations and allow apparently normal differentiation. It is now important to define the epigenetic factors that influence the malignant phenotype to help establish therapeutic strategies for cancer patients" (Hochedlinger, Blelloch, Brennan, Yamada, Kim, Chin and Jaenisch 2004).

In the light of this theoretical perspective it would be wise to overtly recognize that there will not be "the cure" of cancer, as cancer is not a single disease. There certainly will be many small successes that will steadily reduce the overall death rates from various types of cancer, including the invention of strategies that will exploit body's inherent capacity to prevent the growth of the in situ tumors naturally developing along organisms' lifetimes (Folkman and Kalluri 2004).

2.7 Epilogue

The small lesson to take home from the history of 20th Century oncology could be illustrated by slightly modifying the famous dictum of Theodosius Dobzhansky: "nothing in biology [and medicine] makes sense except in the light of evolution[ary thinking]" (Dobzhansky 1971).

References

Anderson, G. R. (2001) Genomic instability and cancer. Current Science. 81, 501–507.
Armitage, P. and Doll, R. (1954) The Age Distribution of Cancer and a Multistage Theory of Carcinogenesis. British Journal of Cancer. 8, 9.
Balmain, A. (2001) Cancer genetics: from Boveri and Mendel to microarrays. Nature Reviews Cancer 1, 77–82.

Berenbluem, I. and Shubik, P. (1949) An Experimental Study of the Initiating Stage of Carcinogenesis, and a Re-Examination of the Somatic Cell Mutation Theory of Cancer. British Journal of Cancer. 3, 109–18.

Berenblum, I. and Shubik, P. (1947) A New, Quantitative Approach to the Study of the Stages of Chemical Carcinogenesis in the Mouse's Skin. British Journal of Cancer 1, 383–91.

Boveri, T. H. (1914) *Zur Frage der Enstehung maligner Tumoren*. Gustav Fisher, Jena.

Breivik, J. and Gaudernack, G. (1999) Carcinogenesis and natural selection: a new perspective to the genetics and epigenetics of colorectal cancer. Adv. Cancer Research 76, 187–212.

Burdette, W. J. (1955) The Significance of Mutation in Relation to the Origin of Tumors: A Review. Cancer Res 15, 201–226.

Cahill, D. P.; Kinzler, K.W.; Vogelstein, B., and Lengauer, C. (1999) Genetic instability and Darwinian selection in tumors. Trends Cell Biol. 12, M57–60.

Comings, D. E. (1973) A General Theory of Carcinogenesis. Proc. Natl. Acad. Sci. 70, 3324–28.

Corbellini G. (2007) EBM. Evolution Based Medicine. Il darwinismo nelle scienze biomediche. Laterza, Roma-Bari.

Dobzhansky, T. (1973) Nothing in Biology Makes Sense Except in the Light of Evolution. The American Biology Teacher 35, 125–129.

Drews, J. (2006) Case histories, magic bullets and the state of drug discovery. Nature Rev. Drug Discovery 5, 635–640.

Fearon, E. R. and Vogelstein, B. (1990) A genetic model for colorectal tumorigenesis. Cell 61, 759–767.

Ferrara, N.; Hillan, K. J.; Gerber, H. P. and Novotny, W. (2004) Discovery and development of bevacizumab, and anti-VEGF antibody for treating cancer. Nature Rev. Drug Discovery 3, 391–400.

Folkman, J. and Kalluri, R. (2004) Cancer without disease. Nature 427, 787.

Foulds, L. (1949) Mammary tumors in hybrid mice; growth and progression of spontaneous tumors. British Journal of Cancer 3, 345–75.

Foulds, L. (1954) The experimental study of tumor progression: a review. Cancer Research 14, 327–339.

Foulds, L. (1956) The histologic analysis of mammary tumors of mice. 2. The histology of responsiveness and progression – The origins of tumors. Journal of the National Cancer Institute 17, 713–756.

Foulds, L. (1958) The natural history of cancer. Journal of Chronic Diseases 8, 2–37.

Foulds, L. (1969) *Neoplastic Development*. Academic Press, New York.

Friend, S. H.; Bernards, R.; Rogelj, S., et al. (1986) A human DNA segment with properties of the gene that predisposes to retinoblastoma and osteosarcoma. Nature 323, 643–46.

Gardner, M. B. (1994) The Virus Cancer Program of the 1970s: a personal and retrospective view. Laboratory Animals Science 44, 101–13.

Hanahan, D. and Weinberg, R. A. (2000) The hallmarks of cancer. Cell 100, 57–70.

Harris, H.; Miller, O. J.; Klein, G.; Worst, P. and, Tachibana, T. (1969) Suppression of malignancy by cell fusion. Nature, 223, 363–8.

Hasegawa, Y.; Ando, Y.; Ando, M.; Hashimoto, N.; Imaizumi, K. and Shimokata, K. (2006) Pharmacogenetic Approach for Cancer Treatment-Tailored Medicine in Practice. Ann. N. Y. Acad. Sci. 1086, 223–232.

Hochedlinger, K.; Blelloch, R.; Brennan, C.; Yamada, Y.; Kim, M.; Chin, L. and Jaenisch, R. (2004) Reprogramming of a melanoma genome by nuclear transplantation. Genes Dev. 18, 1875–1885.

Horrobin, D. F. (2003) Modern biomedical research: an internally self-consistent universe with little contact with medical reality. Nature Rev. Drug Discovery 2, 151–154.

Huebner, R. J. and Todaro, G. J. (1969) Oncogenes of RNA tumor viruses as determinants of Cancer, Proc. Natl. Acad. Sci. 64, 1087–94.

Knudson, A. G. (1971) Mutation and cancer: statistical study of retinoblastoma. Proc. Natl. Acad. Sci. 68, 820–823.

Knudson, A. G. (1971) Mutation and Cancer: Statistical Study of Retinoblastoma. Proc. Nat. Acad. Sci. 68, 820–23.
Knudson, A. G. (1983) Model hereditary cancers of man. Pro. Nucleic Acid Res. Mol. Biol. 29, 17–25.
Knudson, A. G. (1985) Hereditary Cancer, Oncogenes, and Antioncogenes. Cancer Research, 45, 1437–43.
Knudson, A. G. (2005) A personal 60-Year Tour of Genetics and Medicine. Annu. Rev. Genomics Hum. Genet. 6, 1–14.
Koufos, A.; Hansen, M. F.; Copeland, N. G.; Jenkins, N. A.; Lampkin, B. C. and Cavenee, W. K. (1986) Loss of heterozygosity in three embryonal tumors suggests a common pathogenetic mechanism. Nature 316, 330–34.
Leaf, C. (2004) Why we're losing the war on cancer (and how to win it). Fortune 149(6), 76–82, 84–6, 88.
Michor, F.; Iwasa, Y. and Nowak, M. A. (2004) Dynamics of cancer progression. Nature Review Cancer, 4, 197–205.
Michor, F.; Nowak, M. A. and Iwasa, Y. (2006) Evolution of resistance to cancer therapy. Current Pharmaceutical Design 12, 261–271.
Moss, L. (2003) *What Genes Can't Do*. The MIT Press, Cambridge (MA).
Nowell, P. C. (1976) The clonal evolution of tumor cell populations. Science 194, 23–8.
Rous, P. (1943) The Nearer Causes of Cancer JAMA 122, 573–81.
Porter, R. (1997) *The Greatest Benefit to Mankind: A Medical History of Humanity from Antiquity to the Present*. Harper Collins, London.
Schimke, R. T.; Kaufman, R. J.; Alt, F. W. and Kellems, R. F. (1978) Gene amplification and drug resistance in cultured murine cells. Science 202, 1051–5.
Searls, D. B. (2003) Pharmacophylogenomics: genes, evolution and drug targets. Nature Rev. Drug Discovery 2, 613–623.
Smithers, D. W. (1962) An attack on cytologism . The Lancet 1, 493–9.
Stehelin, D.; Varmus, H. E. and Bishop, M. J. (1976) DNA related to the Transforming Gene(s) of Avian Sarcoma Viruses Is Present in Normal Avian DNA. Nature 260, 170–3.
Temin, H. M. (1971) The protovirus hypothesis: speculations on the significance of RNA-directed DNA synthesis for normal development and for carcinogenesis. J Natl Cancer Inst. 46(2), 3–7.
Thagard, P. (1999) *How Scientists Explain Disease*. Princeton University Press, Princeton.
Varmus, H. (2006) The New Era in Cancer Research. Science 312, 1162–1165.
Vineis, P. and Berwick, M. (2006) The population dynamics of cancer: a Darwinian perspective. Int. J. Epidemiol. 35, 1151–9.
Weinberg, R. (2006) *The Biology of Cancer*. Garland Science, New York.
Weiss, L. (2000). Cancer cell heterogeneity. Cancer and Metastasis Reviews 19, 345–350.
Wicha, M. S.; Liu, S. and Dontu, G. (2006). Cancer Stem Cells: An Old Idea – A Paradigm Shift. Cancer Research 66, 1883–1890.

3
Anticancer Drug Discovery and Development

Francesco Colotta, M.D., PhD

3.1 Introduction

The marked contribution of molecular oncology within the past three decades has revealed that the multistage process of cancer growth and progression is attributable to the accumulation of genetic and epigenetic alterations. Malignant carcinomas display genetic alterations in multiple oncogenes and tumor suppressor genes, harbor epigenetic modifications that result in altered expression of several genes and contain chromosomal alterations, including aneuploidy and loss of heterozigosity (Vogelstein and Kinzler 1993; Lengauer, Kinzler, and Vogelstein 1998).

Rapid advances in understanding the molecular mechanisms underlying cancer development and progression should lead to more efficacious and less toxic drugs based on a "rationale" approach to drug discovery in which a new generation of small molecules and monoclonal antibodies are rationally designed to inhibit specific proteins or pathways abnormally expressed in cancer cells. This approach, termed "targeted therapy," is widely applied in the scientific and industrial community to discover new targets and to develop new anticancer drug candidates (Collins and Workman 2006; Weinstein and Joe 2006).

The target-driven drug discovery process is usually depicted as a series of discrete steps for target identification and validation, hit identification and expansion, hit-to-lead and lead optimization, identification of a clinical candidate and clinical studies of Phases I-III. There are minor differences among companies as to what constitutes a "lead" or a "hit," but the general concept is a step-by-step process in which several molecules are synthesized in the early steps, a smaller number of molecules with selected pharmacological properties is characterized further, ending up with (at best) a single candidate molecule worth being tested in humans. This standard process is "hypothesis-driven," with the target as the hypothesis and the drug as the proof of the hypothesis (Sager and Lengauer 2003).

Each step in this process requires the strong integration of many different competences and technologies for optimal progression. The interplay among scientist

---

Vice President, Oncology, Nerviano Medical Sciences, Nerviano (Milan), Italy, Francesco.colotta@nervianoms.com

F. Colotta and A. Mantovani (eds.), *Targeted Therapies in Cancer*.
© Springer 2008

experts in different disciplines – from molecular biology and cell biology to chemistry and Pharmacokinetics and Toxicology, to mention a few – is crucial for a timely and successful drug discovery program.

The aim of this paper is to overview the current small molecule target-driven drug discovery process in Oncology, providing as an example the application of this paradigm in Nerviano Medical Sciences (NMS). NMS (*www.nervianoms.com*) is a research-based company devoted to discover and develop innovative small molecule-based anticancer drugs with emphasis on targeting cell cycle regulating proteins and signal transduction pathways.

3.2 The Evolution of Cancer Drug Discovery

The ultimate goal of any cancer drug discovery process is to discover and develop effective and non-toxic therapies. Over the years this goal has been pursued using approaches driven by the available understanding of the biology of cancer cells and technologies (Suggitt and Bibby 2005).

Over time, a general transition has been observed from the empirical drug screening of cytotoxic agents against poorly characterized tumor models to the target-driven drug screening of inhibitors with a defined mechanism of action. Compounds of both synthetic and natural origin were screened, starting from 1955, in a panel of cancer cell lines and of mice tumors, especially the L1210 leukemia model. The development of nude athymic mice and the successful growth of human tumor xenografts introduced the use of human cancer xenografts in nude mice for screening purposes in 1976. This observation-driven approach has generated cornerstones of cancer therapy, including taxol and antracyclines, with the molecular mechanism of action of these drugs (actually, the target) discovered several years after the identification of the active compound. As an example, the understanding of the molecular mechanism of DNA and DNA-interacting proteins allowed to clarify the mechanism of action of drugs targeting topoisomerases and DNA itself (e.g. etoposide and platinum derivatives) (Suggitt, et al. 2005).

The target-driven approach to drug discovery stems from the current understanding of the molecular mechanism underlying cancer development and progression. The last 25 years of molecular oncology have shown that cancer arises when the right combination of genetic alterations occurs in a susceptible cell. Genetic alterations in cancer involve key regulators of vital cell functions, especially cell cycle and proliferation, apoptosis and cell motility. Proteins aberrantly expressed in cancer cells as a consequence of genetic alterations may represent potential targets for cancer drug discovery. Elucidation of the roles of many kinases, including receptor kinases and signaling kinases, along with the proof that these enzymes are specifically "druggable" targets, prompted the concept of target-driven drug discovery. This "targetcentric" approach to drug discovery can be described as a linear sequence of steps, starting from the identification of a protein altered in cancer cells, followed by the development of an assay assessing the biological activity of the target, screening of compounds inhibiting the target and, after reiterated cycles of optimization and re-testing, identification and selection of inhibitors, with adequate properties (in terms of potency, specificity, drug-like properties, preclinical tolerability) to be tested in animals and in humans for antitumor efficacy and possible toxicological liabilities. (Sager, et al. 2003; Overington, Al-Lazikani and Hopkins 2006; Green 2004; Pegram, Pietras, Bajamonde, Klein and Fyfe 2005).

3.3 Target-Driven Drug Discovery

3.3.1 Target Identification and Validation

Cancer drug discovery is all about making the right decisions. The first crucial decision to be taken is about the target selection. A target should be "validated" at the start of the discovery process and resources should be focused on the most promising ones as early as possible. To this aim, a number of approaches are being used. It is important to underline that none of these approaches should be used in isolation to mark a decision, rather they must be used in combination to select the "right" target (Benson, Chen, Cornell-Kennon, Dorsch, Kim, Leszczyniecka, Sellers and Lengauer 2006; Hooft van Huijsduijen and Rommel 2006).

By and large, three strategies can be used to select a validated target:

- **The genetic approach**: this approach includes the detection of genetic alterations (mutations, amplifications, translocations) in cancer cells' DNA and the differential RNA or protein expression between carefully matched cancer and normal cells or tissues. Today differential expression of transcripts is almost exclusively done by hybridization on chips containing synthetic templates.
- **The functional cell-based approaches**: these include forced expression of an exogenous gene (leading to a transformed phenotype), siRNA and somatic knock-out/knock-in cancer cells. Today siRNA is one of the most popular tools used in target validation.
- **Animal models**: these include mice models (constitutive and inducible transgenic and knock-outs) and model organisms, including xenografts of human tumors. Ideally, the models selected should phanocpy the human disease under investigation.

The kinase cdc7 constitutes an example of a novel/unprecedented target identified and validated in NMS. Cdc7 is Serine/Threonine kinase that was only recently discovered in human cells, based on its similarity with kinases previously characterized in lower organisms (Jiang and Hunter 1997). In these systems, Cdc7 was shown to be essential in promoting the initiation of DNA replication by phosphorylating protein complexes that are bound to DNA sequences that function as replication origins (Bell and Dutta, 2002). DNA replication is a fundamental process for cell proliferation and proteins that participate in this process are conserved throughout the evolution. Importantly mechanisms that control entry in the S-phase and proper execution of DNA synthesis are often altered in malignant cells and at the same time are attractive targets for the development of anticancer agents.

We have used siRNA-mediated down regulation of Cdc7 to determine the function of the kinase in human cells and to understand how normal and tumor cells react to inhibition of Cdc7. We found in a wide panel of cell lines that, consistent with its role in promoting S-phase, Cdc7 depletion causes a marked inhibition of DNA synthesis that correlates with lack of phosphorylation of the Mcm2 sub-unit of the DNA replicative helicase, the Mcm2-7 complex (Fig. 3-1, Panel A). Unlike all current DNA replication inhibitors used in chemotherapy, we observed that Cdc7 inhibition causes selective cell death in tumor cell lines in a p53 independent manner while it causes a reversible cell cycle arrest in normal cells (Montagnoli, Tenca, Sola, Carpani, Brotherton, Albanese, and Santocanale, 2004). Furthermore it does not cause DNA damage, thus not increasing genomic instability, a feature that can lead to therapy resistance. Lack of DNA damage response and induction of apoptosis in cancer cells can easily be assessed by lack of phosphorylation of the Chk1 checkpoint kinase at Ser-345 and by detection of active caspases 3 respectively (Fig. 3-1, Panel B).

Altogether these results indicated that Cdc7 kinase inhibition and the blockade of the initiation reaction is a novel approach for targeting DNA replication compared to other approved DNA , thus providing a rationale for the development of selective Cdc7 kinase inhibitors for the treatment of cancer.

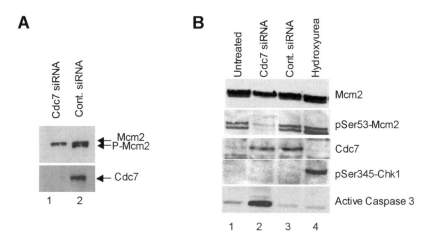

Fig. 3-1 Effects of Cdc7 kinase depletion in HeLa cells. Panel A: Cells were transfected with control (Lane 1) or Cdc7 (Lane 2) siRNA. Protein extracts were prepared and analyzed by Western Blot with anti-Mcm2 and anti-Cdc7 antibodies. Phosphorylation of Mcm2 causes a faster mobility in SDS-PAGE that is lost in Cdc7 depleted cells. Panel B: HeLa cells were either mock treated (Lane 1), transfected with control (Lane 2) or Cdc7 (Lane 3) siRNA or treated with 5 millimolar of ribonucleotide reductase inhibitor Hydroxyurea (Lane 4). Protein samples were prepared and analyzed by Western Blot with the indicated antibodies

3.3.2 Hit Identification

After having identified a promising "good" target, the next step in the drug discovery process is usually to develop a validated assay to test the target biological function and to screen a large compound library (High Throughput Screening, HTS) in order to identify initial "hits," i.e., molecules with limited potency and selectivity in inhibiting the target. Next, after appropriate medicinal chemistry work, hits will be modified and improved to "leads," i.e., molecules with sufficient potency and selectivity to be tested in in vivo animal models. Alternatives to HTS to identify hits exist and will also be discussed hereafter.

3.3.2.1 High Throughput Screening (HTS)

Set up of an appropriate screening strategy is crucial for obtaining solid and reliable results from HTS. In the case of AKT/PKB-directed screening, the Assay Set up Group in NMS designed a specific screening strategy. AKT kinase proteins (AKT1, 2, and 3) are activated by PDK1 kinase through phosphorylation of a conserved threonine residue and act as important signalling elements in the PI3 kinase pathway regulating cell survival and proliferation (Hennessy, Smith, Ram, Lu, and Mills, 2005). An increased level of AKT kinase activity is often found in breast and ovarian tumors and pancreatic and prostate carcinomas and is correlated with a poor prognosis. Inhibition of AKT by antisense RNA reduced tumorigenesis of cancer cells in nude mice. Because of its central role in a highly validated signalling pathway, AKT was chosen as a target for an oncology drug discovery project. A previous HTS campaign already identified very potent and selective competitors for the ATP site of AKTs that bore some cross reactivity towards other kinase targets that could result in unwanted side effects in vivo. Consequently, another HTS campaign was designed to specifically identify non-ATP-competitive inhibitors that would either block the AKT activation step or inhibit AKT through allosteric mechanisms.

This campaign was run against a chemically diverse commercial collection of 200,000 compounds, allowing to profile the type of inhibition of the active compounds already at the primary screening stage.

To identify non-ATP-competitive inhibitors, the compounds were screened against three different kinase assays run in parallel and simultaneously: a PDK/AKT2 coupled assay plus two additional screenings that allowed to de-convolute results and to discard ATP-competitive hits (Fig. 3-2). In the first assay, inactive AKT2 kinase was activated by PDK1, thus becoming capable of phosphorylating a specific substrate not recognized by PDK1. This assay allowed identification of both ATP-competitive and non-competitive inhibitors of PDK1 and AKT2. In the second assay, inhibition of PDK1 phosphorylation of a different substrate was measured, to discriminate inhibitors that directly inhibited PDK1. The third assay used an active form of AKT2 from which the PH domain was deleted.

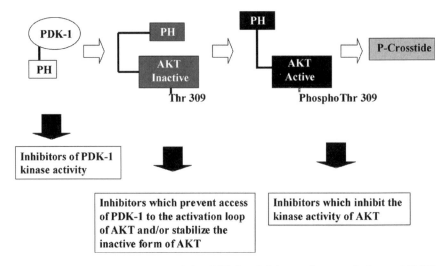

Fig. 3-2 Schematic Representation of the AKT Coupled Assay. Compounds from a 200,000-chemical library were screened against three different kinase assays run in parallel simultaneously. In the first assay, inactive AKT2 kinase was activated by PDK1, thus becoming capable of phosphorylating a specific substrate that was not recognized by PDK1. This assay allowed identification of both ATP-competitive and non-competitive inhibitors of PDK1 and AKT2. In the second assay, inhibition of PDK1 phosphorylation of a different substrate was measured, to discriminate inhibitors that directly inhibited PDK1. The third assay used an active form of AKT2 from which the PH domain was deleted

The μPH domain of AKT2 was chosen with respect to the full-length kinase in order to discriminate PH domain-dependent inhibitors. Positive hits in this assay are expected to be ATP competitors. Each of these assays was designed, optimized and validated using reference inhibitors with the appropriate mechanism of action. In the end, allosteric, non-ATP-competitive and PH-dependent inhibitors could be identified as compounds active in the coupled assay and inactive in the PDK1 and in the ΔPH-AKT2 assays (Table 3-1).

The percent of inhibition in each of the three assays was compared by statistical analysis, which allowed classification of the putative type of inhibition from the primary HTS results. In this way, several hundred unique hits were identified, classified and prioritized for determination of their relative potency and in-depth characterization of the mechanism of inhibition.

After retesting and prioritization of the hits, at least six chemical entities revealed activity towards AKT that was very likely not related to direct ATP competition, suggesting that this approach of coupled assay represents a valuable strategy in the search for non-ATP competitors of kinase targets.

After having set up a suitable screening assay, compound libraries are usually tested by High Throughput Screening (HTS). HTS capabilities in NMS include fully automated compound storage, weighing, dissolution, plating and testing facilities.

3 Anticancer Drug Discovery and Development

Table 3-1 De-convolution of results obtained from HTS of 200,000 compounds in the AKT2 coupled assay. Numbers are nM IC_{50}

Compound	AKT2 ΔPH	PDK1	PDK1/inactive AKT2	Inhibitor
1	114,2	96,3	100,4	Dual AKT2 PDK1
2	82,6	83,5	92,0	Dual AKT2 PDK1
3	139,6	119,8	114,3	Dual AKT2 PDK1
4	−1,0	103,7	90,8	PDK1
5	7,1	112,4	95,5	PDK1
6	2,8	89,5	65,5	PDK1
7	−2,4	76,7	66,8	PDK1
8	3,4	97,1	84,3	PDK1
9	9,5	97,1	90,3	PDK1
10	32,2	119,3	110,8	PDK1
11	3,9	115,4	98,3	PDK1
12	0,2	98,0	83,1	PDK1
13	17,9	117,6	90,4	PDK1
14	77,1	−7,0	75,8	AKT2
15	119,3	9,4	89,3	AKT2
d 16	110,3	−5,6	73,0	AKT2
17	111,3	14,9	69,8	AKT2
18	102,9	9,8	81,7	AKT2
19	95,7	20,2	65,7	AKT2
Staurosporine	103,0	102,2	135,9	Non-specific

Assay formats in NMS include fluorescence, luminescence and radiometric assays in 384 well plates. Two-hundred-thousand compounds can be screened in 10 to 15 days on up to eight different targets simultaneously, generating 1,700,000 experimental points at each screening campaign.

Alternatives to HTS currently used in NMS for Hit Identification are NMR Fragment Screening and Virtual Screening.

3.3.2.2 Nuclear Magnetic Resonance (NMR) Fragment Screening

Fragment-based drug design (FBDD) is a new tool for drug discovery that has emerged in the last decade. Fragments are low molecular weight molecules that interact weakly with a biological target, but display a binding efficiency index (defined as -$LogK_D$/MW or -$LogIC_{50}$/MW where the MW is in KDalton) that is comparable to potent drug molecules. The first step in FBDD is the identification of a suitable fragment that represents a good chemical starting point for optimization of potency, selectivity, pharmacokinetic and pharmaceutical properties. One of the challenges is the reliable detection of the weakly active fragments in the first place.

Nuclear Magnetic Resonance (NMR)-based screening is a suitable methodology for this purpose. Although the absolute sensitivity of the technique is low when compared to other more established techniques used for screening, its relative sensitivity to binding events is second to none. Small molecules binding with affinity in the mM range are easily detected with NMR. Two of these NMR approaches, developed in NMS, utilize the favorable properties of Fluorine NMR spectroscopy. FAXS (Fluorine chemical shift Anisotropy and eXchange for Screening) (Dalvit, Flocco, Veronesi, and Stockman 2002; Dalvit, Fagerness, Hadden, Sarver and Stockman 2003) is a competition-binding assay for the identification of ligands to the target of interest and for the measurement of their dissociation binding constant. 3-FABS (Three Fluorine Atoms for Biochemical Screening) (Dalvit, Ardini, Flocco, Fogliatto, Mongelli, and Veronesi 2003; Dalvit, Ardini, Fogliatto, Mongelli and Veronesi 2004; Dalvit, Mongelli, Papeo, Giordano, Veronesi, Moskau and Kümmerle 2005) is a functional assay for the detection of enzyme inhibitors and for the measurement of the 50 percent mean inhibition concentration (Fig. 3-3).

The robustness of these methodologies, together with the high quality of the generated data allow the selection of molecules displaying only a minute displacement or inhibitory effect thus capturing the broadest chemical structure diversity for potential fragments.

These experiments, coupled with a NMR-based quality control filter, recently developed (Dalvit, Caronni, Mongelli, Veronesi, and Vulpetti 2006), ensure the detection of only **bona fide** ligands and inhibitors avoiding potentially wasted effort in following promiscuous fragments.

3.3.2.3 Virtual Screening

With the increasing number of therapeutic targets, the need for a rapid search for small molecules that may bind to these targets is of crucial importance in the drug discovery process. One way of achieving this is the *in-silico* or virtual screening of large compound collections. If a 3-D structure or model of the biological target is available,

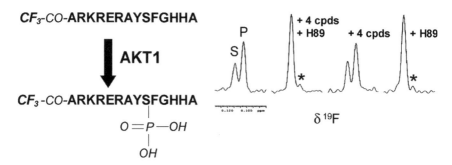

Fig. 3-3 Nuclear Magnetic Resonance fragment screening. Principle of the 3-FABS experiment and its application to the screening and deconvolution of chemical mixtures against the Ser/Thr kinase AKT1. S and P correspond to the ^{19}F NMR signals of the substrate and product of the enzymatic reaction, respectively. The asterisk indicates the small amount of product in the presence of the inhibitor H89. The spectrum on the left is the control sample without test molecules

a commonly used virtual screening technique is high-throughput ligand docking (structure-based virtual screening, SBVS). As a first step, a collection of small molecules is docked into the target structure in a variety of positions and orientations. For each molecule a docking score describing the predicted binding affinity is then calculated on the basis of the complementarity between the docked molecule itself and the biological target in terms of shape and properties such as electrostatics.

A good score indicates that that molecule is potentially a good binder. All molecules in the collection are subsequently rank-ordered by their predicted affinities. This rank-ordered list is then used to select for purchase, synthesis or biological investigation of only those compounds that are predicted to be most active. Typically, this selection will contain a relatively larger proportion of active molecules, i.e., it will be "enriched" with actives compared to a random selection.

Virtual screening has become increasingly important in the context of drug discovery and many successful examples have been reported in the literature (Lyne 2002). In the context of virtual screening high-throughput docking capabilities have been implemented in NMS based on the program QXP distributed onto a Linux cluster. This provides the company with the possibility to screen *in-silico* large collections of drug-like compounds in a relatively small period of time (typically several hundred thousand compounds in a few days). An example of such activities at NMS has recently been published (Trosset, Dalvit, Knapp, Fasolini, Veronesi, Mantegani, Gianellini, Catana, Sundström, Stouten and Moll 2006).

3.3.3 Hit Expansion and Hit-to-Lead

Routinely, a number of hits belonging to different chemical classes are identified by the above mentioned approaches. Hits are usually defined as chemical molecules with suboptimal potency and selectivity in biochemical assays against the intended molecular target. Since it is not uncommon that different chemical classes emerge from HTS, the key decision to be taken at this stage is the selection of a limited number of chemical classes (usually two to three) to be expanded by medicinal chemists. Chemical classes emerging from HTS are usually prioritized on the basis of the evidence of a preliminary Structure Activity Relationship (SAR) in the chemical class, the freedom to operate (intellectual property) and chemical feasibility.

Expansion of hits means that a number of chemical derivatives have to be synthesized starting from the initial hits identified by HTS or other approaches described above. The iterative process of derivative synthesis and testing in the primary biochemical assay (testing the target) and secondary (testing specificity of inhibition) has the objective to design and synthesize more potent and specific chemical compounds with a confirmed mechanism of action and worth being tested in vivo (hit-to-lead). Although there might be a difference among companies as to what constitutes a lead, it is usually considered a lead if a molecule which has a biochemical potency in the low nanomolar range – 10 to 100 fold selectivity on other targets related to the selected target – a potency in cells at least in the low micromolar range, and adequate (although not still optimal) pharmacokinetic properties making it suitable for in vivo testing in animal models of cancer. Initial data on mechanism of action (at least in cells, see below) are also required for a lead.

In order to rank compounds emerging from hit expansion and lead optimization, in NMS we carry out a comprehensive phenotypic profiling of cells exposed in vitro to different concentrations of inhibitors under investigation. Phenotypic profiling in a medium-throughput format includes biochemical readouts related to apoptosis induction, DNA damage response, cell cycle progression/proliferation, signal transduction and histone acetylation. The format is flexible enough to tailor the actual set of readouts to the specific mechanism of action of the target involved.

The medium-throughput of the format allows examining the phenotype induced by several compounds in a dose dependent manner, thus reaching two objectives: firstly, to confirm that the phenotype induced by inhibitors is consistent with the proposed mechanism of action and, secondly, to rank the potency of targeted compounds.

PLK-1 is a serine/threonine protein kinase involved in the entry into, progression through and exit from mitosis, with roles in centrosome maturation, bi-polar spindle formation, chromosome separation and cytokinesis. PLK-1 is ubiquitously expressed in normal tissues and is over-expressed in a wide variety of human tumors; this over-expression correlates with poor prognosis. Several studies have shown that ablation of PLK-1 by either antisense oligonucleotides or siRNA, results in cell growth arrest, mitotic block with aberrant, predominantly monopolar, mitotic spindles and tumor cell death (Strebhardt and Ullrich 2006). Fig. 3-4 shows the high content screening analysis of U2OS cells treated with a PLK-1 inhibitor discovered by NMS scientists. Specific PLK1 inhibitors developed in NMS clearly reproduce PLK1 siRNA results causing a G2/M arrest in asynchronous tumor cells (Fig. 3-4) with monopolar spindle and apoptosis after a prolonged mitotic arrest. Hight Content Screening analysis shows that there is a dose response increase in cyclin B and Histone H3 phosphorilation, in keeping with the notion that PLK-1 inhibition is associated with accumulation of cells in mitosis (Fig. 3-4).

3.3.3.1 Structural Chemistry

Hit expansion and hit-to-lead may take advantage from structural data describing the molecular interactions between the target and hits. To this aim, medicinal chemists work in close contact with computational and structural chemists. Protein X-ray crystallography and molecular modelling are well established as important tools in modern drug discovery. Structural elucidation of a target protein either alone or in complex with small molecule inhibitors provides medicinal chemists with invaluable information for the design of new compounds. In recent years, protein kinases have emerged as potential targets in various therapeutic areas. Thus, these tools have been extensively applied in the search for potent and selective kinase inhibitors.

The vast majority of targets in NMS's pipeline have been successfully crystallized and the structure of targets in association with several small molecule

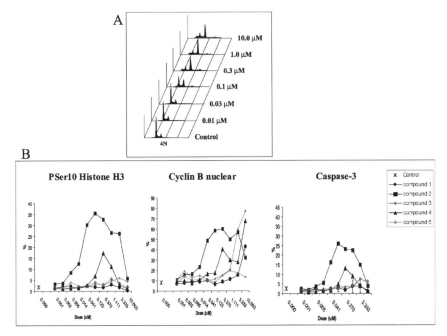

Fig. 3-4 Phenotypic profiling of PLK-1 small molecule inhibitors. Panel A: FACS profile of exponentially growing A2780 cells treated for 24 hours with NMS PLK inhibitor (concentration range 10 nM-10 mM). Accumulation of 4N DNA content cells observed indicative of G2/M block. Panel B: High content cellular assays: multiple mitotic marker and activation of apoptosis could be evaluated in parallel. Five compounds were analyzed in parallel for cyclin B, phospho Ser10Histone H3, and Caspase 3 increases. Compound 2 showing accumulation in mitosis (pSer10Histone H3 and nuclear Cyclin B increase) and apoptosis induction (active Caspase-3 increase) was selected for further evaluation

inhibitors has been solved. Structural data of targets co-crystallized with inhibitors has been instrumental in guiding medicinal chemists into the chemical modifications to be introduced in order to improve potency and selectivity of hits. Moreover, in certain circumstances, structural data have revealed novel and unexpected modes of binding of the inhibitors. This was the case for certain inhibitors of IGF-1R discovered in NMS.

Type I Insulin-like growth factor (IGF-I), IGF II and insulin activate the IGF-I receptor (IGF-IR), which is a transmembrane receptor tyrosine kinase. Several lines of evidences suggest IGF-IR might be involved in cancer growth and progression. High circulating levels of IGF-I are associated with increased risk of developing breast, prostate, colon and lung cancer. In particular, high levels of IGF-I have also been described in several cancers, including breast cancer and multiple myeloma. Finally, IGF-IR stimulates survival, proliferation and motility of different cancer cells (Sachdev and Yee 2007).

In our search for small molecule IGF-1R kinase inhibitors, the pyrrolo[2,3-b] pyridine PHA-552739 was identified as a hit from HTS of our chemical collection.

This molecule was one in a series of compounds that were synthesized for two other kinase projects, and analogs of PHA-552739 had been co-crystallized in-house with these protein kinases, and found to bind within the ATP pocket of their respective target proteins. To investigate the potential of the class as IGF-1R inhibitor, a series of PHA-552739 analogs were designed, and in the (then) absence of a known IGF-1R crystal structure, their binding to the ATP pocket was predicted using an IGF-1R homology model based on the crystal structure of the highly related insulin receptor kinase. However, none of the subsequently synthesised compounds within this series proved superior to PHA-552739 when tested in biochemical assay and the observed Structure-Activity Relationship (SAR) remained flat.

Thus, to better understand the SAR, attempts were made to obtain the crystallographic structure of PHA-552739 in complex with the IGF-1R kinase domain. Eventually, after several co-crystallization experiments conducted with differently phosphorylated forms of the IGF-1R kinase domain, diffraction quality crystals were obtained of PHA-552739 in complex with unphosphorylated IGF-1R (Fig. 3-5). The crystallographic structure revealed that the binding mode of PHA-552739 was different from that observed for the analogs co-crystallized with other protein kinases and, in addition, was unique in comparison to other crystallographically characterized IGF-1R inhibitors. This result explained the observed SAR, and suggested that

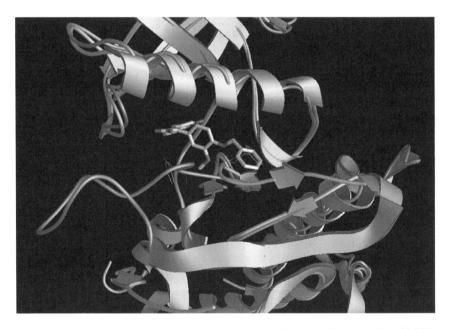

Fig. 3-5 Ribbon diagram of the unphosphorylated IGF-1R kinase domain in complex with PHA-552739. The inhibitor (center) binds into the ATP binding pocket, which is located between the N-terminal (top) and the C-terminal (bottom) lobes of the IGF-1R kinase domain. The pyrrolo (2,3-b)pyridine ring system (pointing towards the back) makes hydrogen bond interactions with the hinge region

PHA-552739 and analogs could represent a new class of compounds able to preferentially interact with the catalytically inactive form of IGF-1R kinase.

3.3.3.2 Biomarker Identification and Validation

During the hit Expansion and lead Optimization stages it is crucial to select inhibitors with a strong correlation between efficacy (in vitro and in vivo) and the studied mechanism of action. Ideally, the compounds to be pushed in the clinic should have demonstrated a good correlation between biochemical and cellular potency, in vivo levels of exposure, efficacy in vivo and modulation of the target in vivo.

To this aim, it is important to identify early in the drug discovery process a biomarker of target modulation. Ideally, the biomarker should be selectively modulated in response to the selected target modulation, and should be detectable in cells and in tissues. Moreover, the selected biomarker should also be suitable for assessing target modulation in human cells, either in surrogate tissues (i.e., skin or blood cells) or in cancer cells from biopsies, in order to make PK/PD correlations in humans. Throughout the drug development process, it is fundamental to apply molecular approaches that allow putting in relationship the efficacy, selectivity and side effect of the compound with its mechanism of action. In particular, during the initial phases of the clinical development, it is important to understand if the target is modulated by the pharmacological treatment, and to correlate this modulation with PK and efficacy data.

The molecular knowledge about the different pathways affected in cancer and the role of key players in these processes is at the basis of the development of targeted drugs. Such knowledge is an integral component of the drug development process, from the choice of the target and the generation of the therapeutic hypothesis about the molecular consequences of its inhibition, to the selection of the most appropriate patients and the monitoring of their response to therapy.

3.3.3.2.1 Biomarkers of Cdc7 Inhibition

When we engaged in the Cdc7 inhibition project, there was no validated biomarker to assess Cdc7 modulation. In order to support a Cdc7 inhibitor drug discovery program, we took a biochemical approach for the identification of specific biomarkers.

The Mcm2 protein is an abundant nuclear protein and its expression is a validated marker of proliferation. As previous experiments indicated that Mcm2 is a good substrate for the Cdc7 kinase (Lei, Kawasaki, Young, Kihara, Sugino and Tye 1997), we used mass spectrometry to identify sites that were specifically phosphorylated by Cdc7 in an **in vitro** reaction. Three serine residues were mapped and antibodies directed against phosphopeptides designed around these residues were raised and characterized. We then observed that in cells Cdc7 inhibition strictly correlated with a decrease of Mcm2 phosphorylation at two of

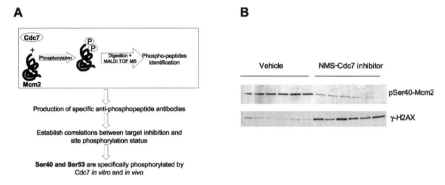

Fig. 3-6 Biomarkers of Cdc7 inhibition. Panel A: Experimental scheme for the identification of Cdc7 specific biomarkers (see text for details). Panel B: Marker modulation in tumors. A2780 xenografts were treated with either vehicle or with a Cdc7 inhibitor. Tumors were explanted and analyzed by Western Blot with anti-pSer40 Mcm2 antibodies or anti-pSer139 H2AX (g-H2AX) antibodies as marker of replication stress and apoptosis. Each lane corresponds to a different tumor

these sites – Ser40 and Ser53 (Fig. 3-6, Panel A) – while at least one other cellular kinase is also involved in Mcm2 phosphorylation in the third site mapped in the **in vitro** experiments (Montagnoli, Valsasina, Brotherton, Troiani, Rainoldi, Tenca, Molinari and Santocanale 2006).

Anti-phosphorylated Mcm2 antibodies are a unique tool to demonstrate the specific mechanism of action of Cdc7 chemical inhibitors, both in cells and in animal models (Fig. 3-6, Panel B), and will offer the potential to assess target inhibition during clinical development.

We identified several low molecular weight inhibitors of cdc7 with a nano-molar potency in the biochemical assay. Several compounds were found efficacious in inhibiting cell proliferation in in vitro and tumor growth in in vivo models. The P-Ser40 MCM2 was inhibited in a dose dependent manner in cancer cells treated with cdc7 inhibitors. Moreover, the mechanism of action was also confirmed ex vivo in tumors from animals treated with cdc7 inhibitors. The good correlation between biochemical potency, in vitro and in vivo efficacy in inhibiting cancer cell growth and target modulation as assessed by biomarker modulation strongly suggest that the anti-tumor activity of cdc7 inhibitors is indeed attributable to in vivo inhibition of the selected target.

3.3.3.2.2 Tracking CDKs Inhibition

Progression through the cell cycle is promoted by cyclin dependent kinases (CDKs), which, when associated with cyclins, push the cells forward through the cell cycle. CDKs are negatively regulated by CDK inhibitors (CDKIs) (Schwartz and Shah 2005; Shapiro 2006). With clinical evaluation of NMS inhibitors currently ongoing, we have attempted to identify genes and proteins whose expression profile is specifically modified by CDK small molecule inhibitors, in correlation with anti-tumor activity, and which might thus be useful as biomarkers of pharmacological activity.

For this purpose we selected, on the basis of literature data, a panel of 44 genes whose expression is described to be regulated by CDK2 either indirectly through phosphorylation of retinoblastoma protein (pRb) and consequent activation of the E2F transcription factor family, or through direct activation of the NPAT transcription factor. Using data from microarray analysis of A2780 human ovarian cancer cells treated in vitro and in vivo with NMS CDK small molecule inhibitors, we selected additional drug-regulated genes to be added to the literature panel (Fig. 3-7).

Modulation of the final set of genes was studied by quantitative RT-PCR in dose response and time course experiments of treatment with CDK small molecule inhibitors both in vitro and in animal models. We found modulation patterns of these gene signatures to be highly correlated with target pathway inhibition (as indicated by biochemical markers) and anti-proliferative effects of CDK inhibitors.

In order to identify biomarkers that might possibly be followed in surrogate tissues from patients undergoing therapy with the NMS CDK inhibitors in cases

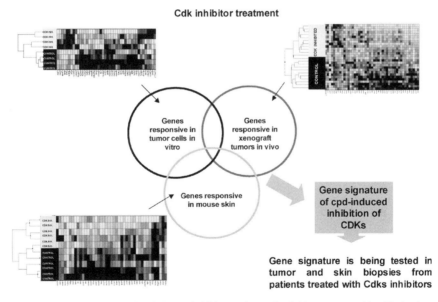

Fig. 3-7 Microarray analysis of CDKs inhibitors. A panel of 44 genes was identified whose expression is described to be regulated by Cdks either indirectly through phosphorylation of retinoblastoma protein (pRb) and consequent activation of the E2F transcription factor family, or through direct activation of the NPAT transcription factor. Modulation of genes was extensively studied by quantitative RT-PCR in dose response and time course experiments of treatment with Cdk inhibitors both in vitro and in animal models. We also analyzed the gene signatures in surrogate tissues such as the skin of mice treated with the compounds. Expression of these gene panels is currently being analyzed in tumor and skin biopsies of patients undergoing treatment with Cdks inhibitors

when acquisition of tumor biopsy is not feasible, we also analyzed the gene signatures in surrogate tissues such as the skin of mice treated with the compounds. Expression of these gene panels is currently being analyzed in tumor and skin biopsies of patients undergoing treatment with CDK small molecule inhibitors in order to support their clinical development and help to understand the feasibility of using gene signature modulation as a biomarker of target inhibition in the clinic (Fig. 3-7).

3.3.4 In Vivo Testing

Compounds with an adequate biochemical and cellular potency and specificity profile and a confirmed mechanism of action in vitro are tested in appropriate in vivo models of cancer growth. Models currently used in NMS include xenografts, transgenic models and target-driven models.

Magnetic Resonance Imaging (MRI) is widely used for the diagnosis and therapy of human tumors in both the clinical and experimental setting. Very similar methods of assessment can be applied to human and animal models, making MRI a unique tool in pre-clinical studies (Lewis, Achilefu, Garbow, Laforest and Welch 2002). By using a dedicated system for small animals (Bruker Pharmascan 7.0 T) we characterize solid tumors **in vivo**, follow their progression and monitor response to therapeutic intervention. By weekly monitoring the animals, MRI is used to detect tumors, monitor cancer progression and evaluate therapeutic treatment.

In the TRAMP model the rat probasin promoter directs expression of SV40 large T antigen to the prostate epithelium leading to abrogation of p53, Rb and PP2A functions. The TRAMP model exhibits progressive stages of prostate cancer ranging from mild to severe prostatic hyperplasia to focal adenocarcinoma and seminal vesicle invasion (Degrassi, Russo, Scanziani, Giusti, Texido, Ceruti and Pesenti 2006) (Fig. 3-8). In our lab we scanned more than 200 mice and a comprehensive study by MRI of pathology progression in this transgenic model was obtained. Different grade tumors were recognized on MR images and **in vivo** MRI findings were validated against classical histological examination.

In the Ki-ras Lat2 model (Johnson, Mercer, Greenbaum, Bronson, Crowley, Tuveson, and Jacks, 2001) the expression of an activated form of the oncogene Ki-ras in the lungs leads to the development of multiple adenocarcinomas (Fig. 3-9). Longitudinal studies by MRI allowed non-invasive monitoring of disease progression through the count of number of lesions and single lesion volume measurement. MRI data were validated versus gross examination and histopathology.

MRI methods allow repeated measurements in transgenic animals and facilitate the translation of findings from pre-clinical to clinical studies. Furthermore, the potential of following tumor response and relapse in single animals adds further value to this methodology that allows the assessment of biological diversity and the investigation of tumor responsiveness and resistance.

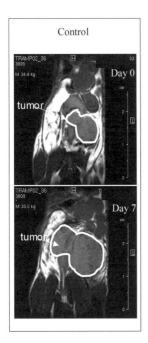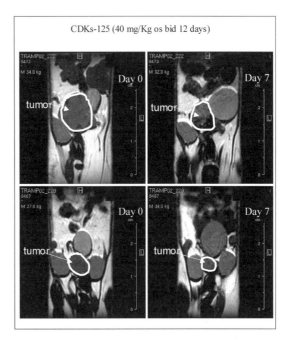

Fig. 3-8 Magnetic Resonance Imaging of the TRAMP tumor model. Rats were untreated (left panels) or treated with a CDK inhibitor at 40 mg/kg bid p. os. (right panels). Rats were then evaluated for prostate cancer development by MRI at different days from the start of the treatment as indicated. Two representative rats are shown in the CDK inhibitor treated group

3.3.5 Clinical Development and Regulatory Affairs Contribution to the Drug Discovery Process

Clinical Development and Regulatory Affairs should contribute to the drug discovery process well before initiation of clinical studies. Their contribution is vital to an early and clear assessment of the medical need and regulatory constrains of each project, thus focusing activities on the critical path, including a feasible clinical plan.

The concept of target-driven drug discovery impacts not only the preclinical stages of the drug discovery process, but also clinical development methods and strategies. Traditionally, anticancer drugs have been tested in patients according to a well defined path of Phase I, Phase II and Phase III studies. Typically, the main objective of phase I studies is to assess the drug safety and to define the dose to be used in phase II studies. From the methodological point of view, in Phase I patients are treated with a dose deemed safe on the basis of animal studies and next the dose is escalated until a toxic dose is reached. The dose just below the toxic one is defined as the maximum tolerated dose (MTD), which is then used in Phase II studies to assess the efficacy of the novel agent by determining the response rate – i.e., the

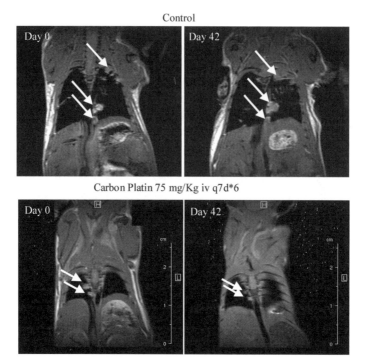

Fig. 3-9 Magnetic Resonance Imaging of the Ki-ras Lat2 model. Rats were untreated (left panel) or treated with Carbon Platin at 75 mg/kg, as indicated. Rats were the evaluated by MRI for the presence of lung adenorcarcinomas at the start of the treatment (Day 0) or after 42 days, as indicated

percentage of patients with a reduction of tumor mass. Finally, Phase III studies are larger and randomized studies intended to compare the novel agent to standard treatments in terms of, typically, patient survival. Phase III studies are required for approval by regulatory authorities.

The concept of target-driven drug discovery challenges the traditional approach to clinical development. The logic consequence of the target-driven approach is that modulation of the target should be the key driver to select doses administered to patients and exposure levels of patients to the new agent. This was indeed the case of Imatinib, in which the MTD was not entirely defined and rather the target was shown to be inhibited in peripheral blood cells at clinically effective doses. However, the logic and appealing concept of defining the biological effective dose as a key decision driver in clinical studies of targeted drugs has technical challenges posed by the availability of tumor samples and the availability of validated methods in patients. The issue of accessibility of repeated tumor samples to measure target modulation might be overcome by molecular imaging and the use of surrogate tissues. An example along this direction is the assessment of Aurora modulation in skin biopsies of patients in Phase I clinical studies with the NMS inhibitor of

Aurora (see paper by Jurgen Moll in this book). However, the use of surrogate tissues must be carefully validated with respect to target modulation in cancer cells.

A key issue raised by the concept of targeted drugs in conducting clinical studies is the relevance of response rate (i.e., tumor shrinkage) in Phase II clinical studies. This is especially compelling because in preclinical studies many targeted drugs show cytostatic rather than cytotoxic activity. Thus, it seems logical to expand the scope of Phase II studies to disease stabilization assessing the rate of progression. However, the rate of progression is considerably variable among cancers, thus making it mandatory to include a control group.

One of the most exciting promises of target-driven drug development was the possibility to select patients who are most likely to respond. Classical inclusion criteria for oncology clinical studies are based on anatomy and histology.

Once again, it seems logical that for targeted drugs patients should rather be selected on the basis of the genotype and the expression of the relevant target. With the usual exceptions (CML and GIST), it has been difficult to establish a correlation between target expression and the response to the targeted drug, and the genotype in solid tumors varies among cells of the same tumor because of genetic instability. As an example, there is no obvious correlation between VEGF expression and response to anti-VEGF antibodies. The difficulty to select patients clearly prevents the clinical development path of targeted drugs from being faster, cheaper and, most importantly, more successful than the traditional path.

3.4 R&D Pipeline in NMS

NMS is the largest R&D site in Italy and one of the largest in Europe focused on Oncology. All competences and technologies required for cancer drug discovery, from molecular biology to cell biology, synthetic and structural chemistry, clinical development and regulatory activities, are represented in NMS. The Project Management group in NMS coordinates the different competencies in order to streamline the entire process of drug discovery. Moreover, the internal Patent Group directly manages more than 600 patents and patent applications, and supports scientists in the definition of patent opportunities.

The current pipeline of targets at NMS includes growth factor receptors, signal transducers, regulators of cell cycle, DNA interacting proteins and molecular chaperones (Fig. 3-10). As of March 2007, NMS project pipeline counts some 12 targets in preclinical stage, with three of them close to the identification of a clinical candidate. Moreover, three new targeted drugs (Aurora and CDK inhibitors) are in phase I and Phase II clinical studies, both in solid and hematological malignancies, whereas two "targeted cytotoxics" (a minor groove binder and a Topoisomerase I inhibitor) are in Phase II clinical studies. The Clinical Development and Regulatory Affairs Departments in NMS are presently involved in 26 clinical studies in more than 60 clinical centers in U.S.and Europe with five new proprietary chemical entities.

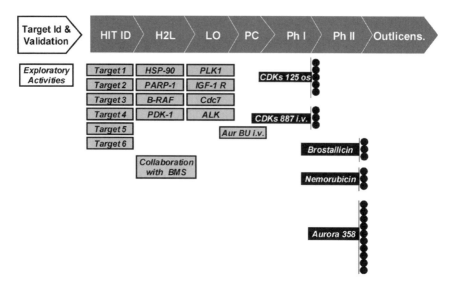

Fig. 3-10 R&D pipeline in NMS. Representation of projects currently under investigation in NMS. Projects are shown at the current different stages of drug discovery process, from preclinical to clinical. For projects in clinical development, the number of clinical studies currently ongoing for each drug are also shown as dots. Projects are considered for out-licensing after Phase II proof-of-concept clinical studies. Target ID: Target Identification; Hit ID: Hit Identification; H2L: Hit to Lead; LO: Lead Optimization; PC: Product Candidate; Ph I- II: Clinical Phase I and II

The NMS project pipeline is especially focused on regulators of cell cycle and signal transducers which are abnormally expressed in cancer cells (Shapiro 2006; Jackson, Patrick, Dar and Huang 2007; Collins and Workman 2006; Sebolt-Leopold and English 2006). Cell cycle represents an ordered and integrated sequence of events that allow cells to grow and proliferate. Several regulators of cell cycle have been described, and many of them are known to be altered in cancer, thus representing potential targets for anticancer drugs. Cell cycle is considered a valid biological target in cancer therapy inasmuch as many drugs currently used in cancer therapy are indeed targeted to regulators of cell cycle.

Most, but not all, targets in NMS are protein kinases (PKs) (Fabbro, Parkinson and Matter 2002; Fancelli and Moll 2005; Krause and Van Etten 2005). The kinome contains some 500 PKs. It is now well established that PKs play a crucial role in several cellular functions, including regulation of cell proliferation and apoptosis. In several cancers, activity of diverse PKs was found to be up-regulated by gain-of-function mutations or over-expression. PKs represent one of the most popular target families in oncology. Targeting of the ATP-binding site of PKs is an attractive and, apparently, simple approach to design small molecule competitive inhibitors of PKs. This approach has led to six low molecular weight kinase inhibitors approved in the clinic (as of Spring 2007), but faces two major issues in the drug development process: entry into the cells accessing the intracellular target and selectivity. The structure-based approach confirms that the ATP-binding domain is conserved among PKs, but the conformation of kinase regions around the ATP-binding site allows for identification of some key diversity which can be exploited to design selective compounds.

Over the years, a competitive technology platform aimed at identifying kinase small molecule inhibitors has been implemented in NMS. Key components of the kinase technology platform in NMS include:

- kinase targeted compound library (KTL). This library includes some 60,000 small chemical molecules rationally designed to perform as ATP-competitive inhibitors in the ATP-binding pocket of PKs
- more than 60 kinases successfully expressed in an enzymatically active conformation and biochemical assays to test enzyme activity have been set up
- different formats are currently used, including fluorescent, luminescent and radiometric detection
- a kinase selectivity screening (KSS) panel in which, thanks to a fully automated and robotized facility, several kinases are tested in parallel for each compound under comparable experimental conditions, thus obtaining a reliable assessment of the selectivity of each compound under evaluation
- HTS automated facility (see above)
- crystallographic and structural data on several kinases, either alone or in association with hundreds of different compounds for each kinase

In conclusion, NMS represents a unique example of an integrated R&D site entirely devoted to small molecule cancer drug discovery with more than 500 scientists.

3.5 Conclusions

The current understanding of the molecular events underlying cancer has profoundly changed the drug discovery process. Over the years, there has been a progressive shift from an approach based on empiricism and observation to a target-driven approach in which drugs are rationally designed to hit and inhibit a specific protein aberrantly expressed in cancer cells believed to play a key role in cancer development and progression. In cancer, kinases have been shown to be the most successful druggable family of proteins to design rational therapies.

The target-driven approach in cancer drug discovery can be described as a sequence of decisions to be taken, from identification of a validated target, to selection of hits and then leads which inhibit the selected target and testing of potent and specific inhibitors in vitro and in vivo models of cancer growth. An exemplification of this paradigm is the drug discovery process in Nerviano Medical Sciences, the largest R&D site in Italy and one of the largest in Europe focused on cancer, in which all competences required for modern drug discovery – from molecular and cellular biology, to medicinal and structural chemistry and clinical development – are represented.

The target-driven approach to drug discovery has generated spectacular improvements in the therapy of cancer, as exemplified by Gleevec and Herceptin. However, the scientific community in academia and industry has to face significant challenges associated with this approach in drug discovery (Kamb, Wee, and Lengauer 2006; Nature Biotech 2005).

The concept of target-driven drug discovery is likely to be especially successful in those cancers in which a molecular drive can be clearly identified. However, most solid tumors do not show a single molecular drive, but rather accumulate a number of genetic alterations leading to the aberrant expression and function of several key regulators of cell proliferation, apoptosis and invasion. An additional obstacle we are facing in applying the concept of targeted therapy is the emergence of drug resistance, which will unavoidably require developing additional agents to be selected on the basis of the molecular mechanisms of resistance.

Although the target-driven approach is currently the most widely accepted paradigm in industry and academia for cancer drug discovery, this approach has proven to be long, expensive and poorly productive (Kamb, Wee, and Lengauer 2006; Nature Biotech 2005). On average it is estimated that the entire drug development process for an oncological product might require up to 10 to 12 years at a cost of $800 million. New approaches are emerging and are being considered in which the emphasis is put on the biology of cancer cells and, ultimately, the patient rather than a predetermined target. Such approaches – termed biology or system-driven – imply that a drug can (or even should) affect multiple targets, in contrast with the target-driven approach instructed by the one-target, one-molecule, one-disease dogma.

Acknowledgments I wish to thank all people in NMS who contributed the experimental data presented in this paper, especially Jurgen Moll, Arturo Galvani, Antonella Isacchi, Enrico Pesenti, Silvia Comis, Adele Giunta, Nicola Mongelli, Lina Speciale, Maria Antonietta Cervini, Corrado Santocanale, Tiziano Bandiera, Jay Bertrand, Christian Orrenius, Fabio Zuccotto, Claudio Dalvit, Pieter Stouten, Anna Degrassi, Micaela Russo, Gemma Texido, Walter Veronelli, Valter Croci, Rosa Giavarini, Rachele Alzani, Roberta Ceruti, Veronica Patton, Alessia Montagnoli, Barbara Valsasina, Sonia Rainoldi, Sonia Troiani, Ermes Vanotti, Giuseppe Locatelli, Sandra Healy, Roberta Bosotti, Emanuela Scacheri, Laura Giorgini, Francesco Sola, Paolo Cappella, Michele Caruso, Daniele Volpi, Cristina Alli, Mariateresa Garavaglia, Nilla Avanzi, Bicia Saccardo, Gianpaolo Fogliatto, Stephane Paris, Raffaella Grimaldi, Dannica Caronni, Patrizia Giordano, Gianluca Papeo and Marina Veronesi. I also thank Anna Migliazza for having reviewed the manuscript.

References

Bell, S. P. and Dutta, A. (2002) DNA replication in eukaryotic cells. Ann. Rev. Biochem. 71, 333–374.
Benson, J. D., Chen, Y. P., Cornell-Kennon, S. A., Dorsch, M., Kim, S., Leszczyniecka, M., Sellers, W. R. and Lengauer, C. (2006) Validating cancer drug targets. Nature. 441, 451–456.
Collins, I. and Workman, P. (2006) New approaches to molecular cancer therapeutics. Nat. Chem. Biol. 2, 689–700.

Dalvit, C., Flocco, M., Veronesi, M. and Stockman, B. I. (2002) Fluorine-NMR competition binding experiments for high-throughput screening of large compound mixtures. Comb. Chem. & HTS. 5, 605–611.

Dalvit, C., Fagemess, P. E., Hadden, D. T. A., Sarver, R. W. and Stockman, B. I. (2003) Fluorine-NMR experiments for high-throughput screening: theoretical aspects, practical considerations, and range of applicability. J. Am. Chem. Soc. 125, 7696–7703.

Dalvit, C., Ardini, E., Flocco, M., Fogliatto, G. P., Mongelli, N. and Veronesi, M. (2003) A general NMR method for rapid, efficient, and reliable biochemical screening. J. Am. Chem. Soc. 125, 14620–14625.

Dalvit, C., Ardini, E., Fogliatto, G. P., Mongelli, N. and Veronesi, M. (2004) Reliable high-throughput functional screening with 3-FABS. Drug Discov. Today 9, 595–602.

Dalvit, C., Mongelli, N., Papeo, G., Giordano, P., Veronesi, M., Moskau, D. and Kúmmerle, R. (2005) Sensitivity improvement in 19F NMR-based screening experiments: theoretical considerations and experimental applications. J. Am Chem. Soc. 127, 13380–13385.

Dalvit, C., Caronni, D., Mongelli, N., Veronesi, M. and Vulpetti, A. (2006) NMR-based quality control approach for the identification of false positives and false negatives in high throughput screening. Curr. Drug Disc. Tech. 3, 115–124.

Degrassi, A., Russo, M., Scanziani, E., Giusti, A., Texido, G., Ceruti, R. and Pesenti, E. (2006) Magnetic resonance imaging and histopathological characterization of prostate tumors in TRAMP mice as model for preclinical trials. Prostate. 67, 396–404.

Fabbro, D., Parkinson, D. and Matter, A. (2002) Protein tyrosine kinase inhibitors: new treatment modalities? Curr. Opin. Pharmacol. 2, 374–381.

Fancelli, D. and Moll, J. (2005) Inhibitors of Aurora kinases for the treatment of cancer. Expert Opin. Ther. Patents 15, 1169–1182.

Green, M. R. (2004) Targeting targeted therapy. N. Engl. J. Med. 350, 2191–2193.

Hennessy, B. T., Smith D. L., Ram, P. T., Lu, Y. and Mills, G. B. (2005) Exploiting the PI3K/AKT pathway for cancer drug discovery. Nature 4, 988–1004.

Hooft van Huijsduijnen, R. and Rommel, C. (2006) De-compartmentalizing target validation – thinking outside the **pipeline** boxes. J. Mol. Med. 84, 802–813.

Jackson, J. R., Patrick, D. R., Dar, M. M. and Huang, P. S. (2007) Targeted anti-mitotic therapies: can we improve on tubulin agents? Nat. Rev. Cancer 7, 107–117.

Jiang, W. and Hunter, T. (1997) Identification and characterization of a human protein kinase related to budding yeast Cdc7p. Proc. Natl. Acad. Sci. USA 94, 14320–14325.

Johnson, L., Mercer, K., Greenbaum, D., Bronson, R. T., Crowley, D., Tuveson, D. A. and Jacks, T. (2001) Somatic activation of the **K-ras** oncogene causes early onset lung cancer in mice. Nature 410, 1111.

Kamb, A.,Wee, S. and Langauer C. (2007) Why is cancer drug discovery so difficult? Nat. Rev. Drug Discov. 6, 115–120.

Krause, D. S. and Van Etten, R. A. (2005) Tyrosine kinases as targets for cancer therapy. N. Engl. J. Med. 353, 172–187.

Lei, M., Kawasaki, Y., Young, M. R., Kihara, M., Sugino, A. and Tye, B. K. (1997) Mcm2 is a target of regulation by Cdc7-Dbf4 during the initiation of DNA synthesis. Genes Dev. 11, 3365–3374.

Lengauer, C., Kinzler, K. W. and Vogelstein B. (1998) Genetic instabilities in human cancers. Nature 396, 643–649.

Lewis, J. S., Achilefu, S., Garbow, J. R., Laforest, R. and Welch, M. J. (2002) Small animal imaging: current technology and perspectives for oncological imaging. Eur. J. Cancer 38, 2173–2188.

Lyne, P. D. (2002) Structure-based virtual screening: an overview. Drug. Discov. Today 7, 1047–1055.

Montagnoli, A., Tenca, P., Sola, F., Carpani, D., Brotherton, D., Albanese, C. and Santocanale, C. (2004) Cdc7 inhibition reveals a p53-dependent replication checkpoint that is defective in cancer cells. Cancer Res. 64, 7110–7116.

Montagnoli, A., Valsasina, B., Brotherton, D., Troiani, S., Rainoldi, S., Tenca, P., Molinari, A. and Santocanale, C. (2006) Identification of Mcm2 Phosphorylation Sites by S-phase-regulating Kinases. J. Biol. Chem. 281, 10281–10290.

Nature Biotech, (2005) A dose of reality for rational therapies. Nat. Biotech. 23, 267.

Overington, J. P., Al-Lazikani, B. and Hopkins, A. L. (2006) How many drug targets are there? Nat. Rev. Drug Disc. 5, 993–996.

Pegram, M. D., Pietras, R., Bajamonde, A., Klein, P. and Fyfe, G. (2005) Targeted therapy: wave of the future. J. Clin. Oncol. 23, 1776–1781.

Sachdev, D. and Yee, D. (2007) Disrupting insulin-like growth factor signalling as a potential cancer therapy. Mol. Canc. Ther. 6, 1–12.

Sager, J. A. and Lengauer, C. (2003) New paradigms for cancer drug discovery? Canc. Biol. & Ther. 2, 178–181.

Schwartz, G. K. and Shah, M. A. (2005) Targeting the cell cycle: a new approach to cancer therapy. J. Clin. Oncol. 23, 9408–9421.

Sebolt-Leopold, J. S. and English, J. M. (2006) Mechanisms of drug inhibition of signalling molecules. Nature. 441, 457–462.

Shapiro, G. I. (2006) Cyclin-Dependent kinase pathways as targets for cancer treatment. J. Clin. Oncol. 24, 1770–1783.

Strebhardt, K. and Ullrich, A. (2006) Targeting polo-kinase 1 for cancer therapy. Nat. Rev. Cancer. 6, 321–330.

Suggitt, M. and Bibby, M. C. (2005) Fifty years of preclinical anticancer drug screening: empirical to target-driven approaches. Clin. Can. Res. (11) 971–981.

Trosset, J. Y., Dalvit C., Knapp, S., Fasolini M., Veronesi, M., Mantegani S., Gianellini M., Catana C., Sundström M., Stouten P. F. W. and Moll J. K. (2006) Inhibition of protein-protein interactions: the discovery of drug-like beta-catenin inhibitors by combining virtual and biophysical screening. Proteins 64, 60–67.

Vogelstein, B. and Kinzler K. W. (1993) The multi-step nature of cancer. Trends Genet. 9, 138–141.

Weinstein, I. B. and Joe, A. K. (2006) Mechanisms of disease: oncogene addiction – a rationale for molecular targeting in cancer therapy. Nat. Clin. Pract. Onc. 3, 448–457.

4
Beyond VEGF: Targeting Tumor Growth and Angiogenesis via Alternative Mechanisms

James Christensen, Ph. D. and Kenna Anderes Ph. D.

4.1 Introduction

Cancer is a multiplicity of diseases characterized by a range of molecular defects leading to unregulated, aberrant cell growth. Recent improvements in understanding the molecular basis of cancer progression, improved diagnostics, and the emergence of new classes of therapeutics have offered promise to better manage the disease. Chemotherapeutic agents that induce cytotoxicity by damaging DNA have been the mainstay of cancer treatment for many decades. Despite their effectiveness, there are a number of limitations, most notably a narrow therapeutic window due to lack of selectivity toward cancer cells. A new generation of chemotherapeutic agents commonly referred to as "targeted therapies" is emerging and aims to impart selectivity to cancer cells by exploiting molecular differences between normal and cancer cells. The best known examples of these new drugs are imatinib and erlotinib, both acting through inhibition of the tyrosine kinases (TKs) BCR-ABL and EGFR, respectively. Lessons learned from imatinib and erlotinib include the necessity for careful patient selection, (i.e., identification of the underlying molecular anomalies) to optimize the potential efficacy of targeted therapies. In this chapter, we describe two novel kinase targeted therapies with distinct mechanisms of action: checkpoint kinase 1 (Chk1) and c-Met kinase inhibitors. Inhibition of Chk1 represents a molecularly targeted approach to selectively enhance the cytotoxicity of DNA-damaging agents in tumor cells with intrinsic checkpoint defects (mutated p53) while minimizing toxicity in normal cells that have a checkpoint competent molecular phenotype (wild-type p53). In contrast, an extensive body of literature indicates that c-Met is one of the most frequently genetically altered or otherwise dysregulated RTKs in advanced cancers implicating it as a key target for therapeutic intervention.

Department of Cancer Biology, Pfizer Global Research and Development, La Jolla Laboratories, 10724 Science Center Drive, La Jolla, CA 92121, james.christensen@pfizer.com

4.2 Checkpoint 1 (Chk1)

Checkpoints are present in all phases of the cell cycle and are regarded as the gatekeepers maintaining the integrity of the genome (1-3). Many conventional agents used to treat cancer impart damage to the genome and activate cell cycle checkpoints. p53 is the major player in the checkpoint that arrests cells at the G_1/S boundary, while Chk1 is critical for the S and G_2/M checkpoints (4-6). Dysfunction in cell cycle checkpoints appears to be a universal phenomenon in human cancers. The p53 tumor suppressor gene is mutated more often in human cancers than any other gene yet reported (7). These cancers lack a functional G_1 checkpoint; however, the $S/G_2/M$ checkpoints remain intact and, in response to DNA damage, arrest cell cycle progression allowing time for DNA repair. The essential and indispensable role of Chk1 in initiating the $S/G_2/M$-phase checkpoints has been demonstrated through biochemical and genetic studies (8, 9). A recent review summarizes key findings validating Chk1 as a therapeutic target (10). Inhibiting Chk1 represents a therapeutic strategy for creating a "synthetic lethal" response by overriding the tumor cell's last checkpoint defense against the lethal damage induced by DNA directed chemotherapeutic agents. The studies below describe the identification and characterization of PF-00477736, a potent, selective ATP-competitive small molecule inhibitor of Chk1.

4.3 Results and Discussion

PF-00477736 was demonstrated to be a potent ATP-competitive inhibitor of Chk1 kinase activity with a biochemical K_i of 0.49 ± 0.29 nM. PF-00477736 was evaluated across a panel of >100 diverse receptor tyrosine and serine-threonine kinases. PF-00477736 exhibited a high degree of selectivity at pharmacologically relevant concentrations. PF-00477736 significantly inhibited the activity of Chk2 ($K_i = 47 \pm 9$ nM, essentially 100-fold vs. Chk1) and VEGFR2 ($K_i = 8 \pm 1$ nM, 16-fold vs Chk1), but was a poor inhibitor of CDK1 activity ($K_i = 9.9$ uM, 20,000-fold vs. Chk1). PF-00477736 inhibited 6 kinases with less than 100-fold selectivity (based on IC_{50} to K_i ratio): Aurora-A, FGFR3, Flt3, Fms (CSF1R), Ret, and Yes. For PF-00477736, kinases that are most pharmacologically relevant for selectivity considerations are those for which transient intermittent inhibition would influence cell cycle progression (i.e., CDKs, mitotic kinases), checkpoint control (e.g., Chk2, ATM, ATR), or act on apoptotic pathways (e.g., AKT, p38). Based on this, VEGFR2, Fms/CSF1R, FGFR2, Flt3, and Ret are not considered to be relevant because sustained inhibition is required to evoke observable pharmacology from these RTKs. Aurora-A is a relevant kinase, however the enzyme assay did not correlate with cell-based functional activity. In cell-based functional assays, PF-00477736 showed >100-fold selectivity ratio for Chk1 vs Aurora and Chk1 vs Chk2.

Checkpoint abrogation, as measured by an increase in Histone H3 phosphorylation, was demonstrated for PF-00477736 in combination with several DNA-damaging agents including camptothecin, gemcitabine or radiation. PF-00477736 inhibited camptothecin-induced G_2 arrest in p53-mutated human lymphoma CA46 cells (EC_{50} 45 nM). Under similar conditions, high content imaging with an algorithm created to quantify the mitotic index based on phosphohistone H3 expression revealed similar abrogation of camptothecin-induced G2 arrest in HeLa cells (EC_{50} 38 nM). Further characterization by flow cytometry analysis in HT29 cells indicated no change in cell cycle profile for PF-00477736 in the absence of DNA damage, while gemcitabine alone induced a prominent S-Phase arrest. PF-00477736 effectively abrogated gemcitabine-induced S-phase arrest and increased apoptotic cell death 5-fold (Fig. 4-1). In addition, PF-00477736 demonstrated time-dependent modulation of key effectors in the Chk1 signaling pathway. PF-00477736 showed a decrease in the phosphorylation of CDK1, the presence of the mitotic, hyperphosphorylated form of Cdc25C, and an increase in cyclin B levels followed by complete disappearance; all of which collectively provide evidence of checkpoint abrogation, mitotic entry, and cell cycle progression through metaphase of mitosis. Further, PF-00477736 in combination with gemcitabine showed accumulation of -H2AX foci, indicating greater DNA damage presumably as a result of checkpoint abrogation and lack of DNA repair. PF-00477736 enhanced the

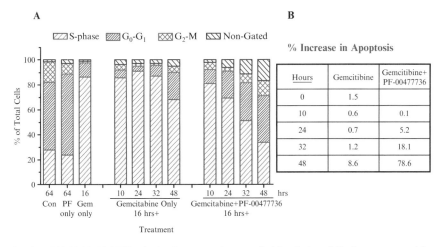

Fig. 4-1 PF-00477736 Effectively abrogates the gemcitabine-induced S-phase arrest with a corresponding increase in apoptotic cell populations in the combination treatment compared to the gemcitabine treatment alone (A) HT29 cells were treated with gemcitabine for 16 hours followed by PF-00477736 at times indicated. Cells were labeled with BrdU prior to harvest. Upon FITC-conjugated anti-BrdU and propidium iodide staining, cells were analyzed using a Becton Dickinson FACSCalibur. The proportion of cells in G_1, S or G_2-M was determined using CellQuest software. (B) HT29 cells were treated as described for (A) and fixed after harvest. Cells were stained with FITC-dUTP and propidium iodide, and analyzed with FACSCalibur. The proportion of cells with DNA strand breaks was determined with CellQuest software

Table 4-1 In vitro cytotoxicity of PF-00477736 in selected cell lines with different DNA-damaging agents. Summary of PF-00477736 **in vitro** cytotoxicity in combination with different DNA-damaging agents in selected cell lines as analyzed by MTT assays

Cell line (Tumor Type)	HT29 (Colon)	Colo205 (Colon)	PC-3 (Prostate)	MDA-MB-231 (Breast)	K562 (Leukemia)
PF-00477736 IC_{50} (uM)	1.8	1.3	1.6	1.4	0.42
PF-00477736 TI^a	8.5	14	13.2	2.1	9.3
DNA damaging agent (Classification)	\multicolumn{5}{c}{PF_{50}^b induced by PF-00477736}				
Gemcitabine (Antimetabolite)	9	11.3	12.2	3.6	5.6
SN-38 (Topo 1 inhibitor)	3.7	2.1	1.3	2.4	1.9
Carboplatin (Alkylating agent, cross linking agent)	3	5.4	3.1	2.55	1.9
Doxorubicin (DNA intercalcator, non-covalent binding)	2.2	1.1	1.5	2.25	1.1
Mytomicin C (Alkylating agent, Aziridine)	3.7	5.3	ND^c	1.2	ND^c

^{a}OTSI (on-Target Selectivity Index) was calculated as IC_{50}, (PF-00477736 alone)/IC_{50}, (combination treatment). $^b PF_{50}$ (Potentiation factor $_{50}$) was calculated as IC_{50}, (cytotoxic agent alone)/IC_{50}, (combination treatment). ^{c}ND: not determined. In these assays the curves' profile did not allow calculation of an accurate PF_{50}.

cytotoxicity of a broad spectrum of clinically important chemotherapeutic agents including, but not limited to, gemcitabine, irinotecan, doxorubicin and carboplatin in a panel of p53-defective human cancer cell lines. PF-00477736 alone caused no significant effect on cell viability compared with untreated cells. A representative example of PF-00477736 chemopotentiation was illustrated by significant enhancement (89 percent) of gemcitabine cytotoxicity compared with gemcitabine alone. PF-00477736 induced robust and consistent potentiation with most agents with some variability observed between cell lines (Table 4-1).

The functional status of p53 is material to the checkpoint kinase inhibitor approach to selectively enhance the cytotoxicity of DNA-damaging agents in tumor cells with intrinsic checkpoint defects (mutated p53) while minimizing toxicity in normal p53-competent cells. Cell survival assays were performed in p53 wild-type or mutant HTC116 cells. PF-00477736 combined with gemcitabine induced 44% cell growth inhibition in the mutant p53 HCT116 cells compared to 15 percent cell growth inhibition in p53 wild-type. These results suggest p53-defective cancer cells are more vulnerable to Chk1 inhibition compared to their p53-competent counterparts. PF-00477736 was also evaluated in p53-competent non-tumor HUVEC (human umbilical vein endothelial) cells and demonstrated negligible change in cytotoxicity in combination with either gemcitabine or camptothecin.

PF-00477736 demonstrated dose-dependent potentiation of gemcitabine or irinotecan antitumor efficacy in xenografts models. HT29 and Colo205 human colon carcinoma xenografts were selected as models for combination studies based on their p53-deficient status and chemosensitivity profile to gemcitabine and irinotecan. The gemcitabine or irinotecan doses and schedules examined represent clinically relevant regimens adapted to enable combination studies in mice (11, 12). The optimal dosing schedule for PF-00477736 was determined to be sequential administration with a 24-hour interval between cytotoxic agent and PF-00477736. The rationale for this sequence considers time to be important to allow Chk1-mediated accumulation of S/G_2M-arrested cells rendering them vulnerable to S/G_2M-checkpoint inhibition while also allowing normal cells time to repair damaged DNA. PF-00477736 demonstrated no single-agent antitumor activity in the Colo205 or HT29 xenografts. PF-00477736 dose dependently enhanced the antitumor activity of a maximum tolerated dose of gemcitabine with no concomitant exacerbation of side effects. PF-00477736 (40 mg/kg) enhanced gemcitabine tumor growth inhibition by 75 percent with a time-to-progression enhancement ratio of 50 percent. Log cell kill was increased from 0.5 to 1.25 and hazard ratio analysis indicated the time-to-progression endpoint was improved 3.6-fold, compared with gemcitabine treatment alone. Upon combination with a maximum tolerated dose of irinotecan (100 mg/kg) PF-00477736 significantly potentiated the time-to-progression endpoint demonstrating 2.8-fold greater efficacy compared to irinotecan administered as a single agent. Notably, PF-00477736 in combination with half the maximum tolerated dose of irinotecan (50 mg/kg) achieved efficacy equivalent to that obtained with the full maximum tolerated dose. In all combinations PF-00477736 was well tolerated and demonstrated a significant enhancement in therapeutic index.

To gain confidence that enhanced antitumor activity correlated with modulation of Chk1 activity, tumor sections were evaluated utilizing immunohistochemical methods for Chk1 kinase activation (assessed by phosphorylation on Chk1-Ser317) and apoptosis (assessed by Caspase 3 activation). Significant dose-dependent inhibition of gemcitabine-induced Chk1 phosphorylation (90 percent) was observed for PF-00477736, indicating a reduction of activated Chk1. In addition, a qualitative induction of caspase-3 was observed indicating increased apoptosis. Collectively, these data serve to establish a correlation between antitumor activity and inhibition of Chk1 activity.

In summary, these studies describe the key pharmacologic effects of PF-00477736 in combination with common DNA-directed agents on a variety of tumor cell lines and tumor models. PF-00477736 demonstrated checkpoint abrogation, enhanced cytotoxicity, potentiated antitumor activity and selectivity for p53-mutated phenotype. PF-00477736 combinations were well tolerated with no exacerbation of side effects commonly associated with cytotoxic agents. Key issues in the clinical development of Chk1 inhibitors include defining optimal combinations and dosing regimens for achieving maximal therapeutic window and stratification of patients based on p53 status. The clinical attractiveness of combinations with PF-00477736 includes the potential for enhancing the efficacy of widely used chemotherapeutics, improved therapeutic index and quality of life due to preferential killing of cancer cells.

4.4 c-Met

c-Met is the prototypic member of a unique subfamily of RTKs. It is broadly expressed in a variety of epithelial-derived tissues and, along with its cognate ligand – HGF – is an important regulator of mammalian development, tissue repair, and homeostasis (13, 14). The dysregulation of c-Met and HGF are implicated in metastatic cancer progression. c-Met and HGF are highly expressed relative to surrounding tissue in numerous cancers and their expression correlates with poor patient prognosis (13). Additionally, c-Met activating point mutations or amplification of the c-Met gene locus are implicated in progression of multiple cancer types (15-20). Cell lines engineered to express high levels of c-Met and HGF (autocrine loop) or mutant c-Met displayed a proliferative, motogenic and/or invasive phenotype and grew as metastatic tumors in nude mice (21-25). Furthermore, HGF and c-Met have been implicated in the regulation of tumor angiogenesis through the direct proangiogenic properties of HGF or through regulation of secretion of angiogenic factors including VEGFA, interleukin-8 (IL-8), and thrombospondin-1 (26-29).

Collectively, the aforementioned data indicates that dysregulation of c-Met is a factor in progression of a variety of cancers (30). The studies below describe the identification and characterization of PF-2341066, an orally available ATP-competitive and selective small molecule inhibitor of c-Met.

4.5 Results and Discussion

PF-2341066 was demonstrated to be a potent ATP-competitive inhibitor of c-Met kinase activity with a biochemical K_i of 4 nM and a mean IC_{50} value of 11 nM for inhibition of c-Met phosphorylation across a panel of human cell lines. PF-02341066 also demonstrated similar potency against a variety of mutant variants of c-Met identified in papillary renal carcinoma, head and neck, and lung cancers indicating that an intervention strategy in these tumor types may be effective (15-19).

In addition to c-Met, PF-2341066 was evaluated in >120 biochemical kinase assays representing a structurally diverse cross-section of nearly 25 percent of all known tyrosine and serine-threonine kinases (31). Data indicated that PF-2341066 exhibited a high degree of selectivity at pharmacologically relevant concentrations. The exception to this was ALK which was inhibited at pharmacologically relevant (24 nM) concentrations of PF-2341066 in lymphoma cells expressing the oncogenic NPM-ALK fusion protein. These data indicate that the pharmacologic activity of PF-2341066 is likely mediated by inhibition of c-Met and ALK RTKs and their oncogenic variants. These findings are important in interpreting the pharmacologic properties of PF-2341066 and in determining which patient populations are most likely to exhibit clinical benefit from treatment with this agent. The inhibition of c-Met is likely to be relevant in tumor types in which this target is genetically dysregulated or otherwise abnormally activated and may also be associated with antiangiogenic activity in selected tumor types. Wild-type ALK is not broadly expressed across tumor types, however; ALK fusion proteins are implicated in the pathogenesis of anaplastic large cell lymphoma (ALCL) and inflammatory myofibroblastic tumors (IMT) (32).

Because c-Met is implicated in a variety of tumor cell and tumor endothelial cell functions, PF-2341066 was evaluated in a series of cell-based functional assays. PF-2341066 inhibited human GTL-16 gastric carcinoma cell growth (IC_{50} = 9.7 nM), induced apoptosis in GTL-16 cells (IC_{50} = 8.4 nM), inhibited HGF-stimulated human NCI-H441 lung carcinoma cell migration and invasion (IC_{50s} of 11 nM and 6.1 nM, respectively), and inhibited MDCK cell scattering (IC_{50}: 16 nM). In studies performed utilizing Human Umbilical Vascular Endothelial Cells (HUVEC), PF-2341066 inhibited HGF-stimulated c-Met phosphorylation (IC_{50}: 11 nM), cell survival (IC_{50}: 14 nM), and matrigel invasion (IC_{50}: 35 nM) as well as serum-stimulated tubulogenesis (formation of vascular tubes) of Human Microvascular Endothelial Cells (HMVEC). The strong correlation of IC_{50} values for inhibition of c-Met phosphorylation and c-Met-dependent phenotypes suggest that PF-2341066 pharmacologic activity in these assays is mediated by inhibition of c-Met. In addition, this data indicates that potential anticancer effects of PF-2341066 may range from 1) direct effects on tumor cell growth and survival to 2) antimetastatic properties through inhibition of tumor cell migration and invasion and may include 3) potential antiangiogenic activity mediated through direct effects on ability of endothelial cells to form functional vasculature.

To evaluate the ability of PF-2341066 to inhibit c-Met in tumor in vivo, GTL-16 gastric carcinoma or other tumors were harvested at several time points following its oral administration over a range of doses to assess c-Met phosphorylation by ELISA (Fig. 4-2). The following conclusions were apparent from these studies: 1) complete inhibition of c-Met activity for 24 hours is consistent with complete inhibition of tumor growth (50 mg/kg); 2) potent inhibition of c-Met activity for only a portion of the schedule is consistent with significant, but not complete efficacy (12.5 mg/kg, 60 percent TGI); 3) inability to achieve >50 percent inhibition of c-Met activity (3.125, 6.25 mg/kg) is consistent with lack of significant tumor growth inhibition (TGI). Collectively, these studies suggest that near-complete inhibition of c-Met phosphorylation (>90 percent inhibition) for the duration of the administration schedule is necessary to maximize antitumor efficacy of PF-2341066. In efficacy studies, PF-2341066 was well tolerated at the maximally efficacious dose of 50 mg/kg/day for as long as three months or at 200 mg/kg/day for as long as one month with no weight loss or evidence of overt toxicity or histopathological findings.

The efficacy of PF-2341066 was further assessed at optimal dose levels (i.e., 50 mg/kg/day) in human tumor xenograft models representative of cancer indications in which dysregulation of c-Met is implicated. In these studies, PF-2341066 demonstrated cytoreductive antitumor activity in models expressing high levels of constitutively activated c-Met including GTL-16 gastric (60 percent decrease in mean tumor volume) or Caki-1 renal carcinoma (53 percent decrease in mean

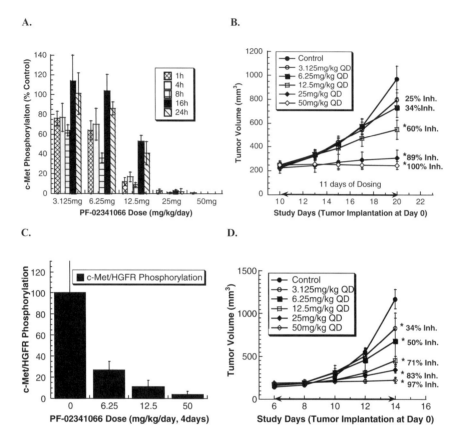

Fig. 4-2 Inhibition of c-Met phosphorylation (A) and tumor growth (B) by PF-2341066 in GTL-16 xenograft model Athymic mice bearing established GTL-16 (250 mm^3) (A and B) or U87MG (180 mm^3) (C and D) tumors were administered PF-2341066 orally at the indicated dose or vehicle alone over the designated treatment schedule. For studies investigating inhibition of c-Met phosphorylation (A or C), mice were humanely euthanized at designated time points post-administration, tumors were resected and frozen, and phosphorylation in vehicle and treated groups was quantified by ELISA. Inhibition of kinase target phosphorylation by PF-2341066 in tumors was calculated as: % Inhibition = 100-[(Mean OD treated / Mean OD untreated) X 100]. For studies investigating tumor growth inhibition (B or D), tumor volume was measured using Vernier calipers on the indicated days with the median tumor volume ± SEM indicated for groups of 15 mice. Percent (%) tumor growth inhibition values measured on the final day of study for drug-treated compared to vehicle-treated mice and are calculated as: $100^* \{1-[(\text{Treated}_{\text{Day 20}} - \text{Treated}_{\text{Day 10}}) / (\text{Control}_{\text{Day 20}} - \text{Control}_{\text{Day 10}})]\}$. An * denotes that the median tumor volumes are significantly less in the treated vs. the control group (P ≤ 0.001) as determined using one-way analysis of variance

tumor volume). This data indicates that gastric carcinoma patients with amplified c-Met or renal patients with high levels of constitutively activated c-Met may comprise clinical populations that exhibit higher probability of objective response to PF-2341066. PF-2341066 also demonstrated near complete inhibition of the growth of established U87MG glioblastoma or PC-3 prostate carcinoma xenografts with 97 perecent or 84 percent inhibition on the final study day, respectively, supporting the concept that tumor types which exhibit autocrine production of c-Met and HGF may also be candidates for clinical studies. In contrast, although a robust response of the NCI-H441 NSCLC paracrine model (43 percent decrease in mean tumor volume) to PF-23410666 was observed, other paracrine c-Met expressing models such as MDA-MB-231 breast and DLD-1 colon carcinoma did not respond. This data suggests that response of populations in which wild-type c-Met is expressed in absence of the constitutively active RTK or autocrine HGF may exhibit a heterogeneous response to c-Met-directed therapy. Further study of molecular determinants sensitivity to c-Met –directed therapy is warranted.

To further understand the mechanism of PF-2341066 efficacy, it was evaluated for its effect on tumor mitotic index (Ki67), apoptosis (activated caspase-3), and microvessel density (MVD by CD-31) utilizing immunohistochemical (IHC) methods. A significant 4-fold decrease in Ki67 levels by study day 4 was observed at 25 and 50 mg/kg/day PF-2341066 in the GTL-16 and U87MG models indicating a correlation with maximum antitumor efficacy. Similarly, a dose-dependent induction of activated caspase-3 at 25 and 50 mg/kg/day was observed in the GTL-16 model and at 50 mg/kg/day in the U87MG model. Therefore, one mechanism of PF-2341066 is likely mediated via direct effects on tumor cell mitogenesis and apoptosis in tumor types in which dysregulation of c-Met is implicated in progression.

To assess antiangiogenic activity of PF-2341066, a significant dose-dependent reduction of CD31 positive endothelial cells was observed in GTL-16 tumors by study day 11 indicating that inhibition of MVD correlated to antitumor efficacy. In addition, PF-2341066 demonstrated a significant dose-dependent reduction of plasma levels of the human proangiogenic factors, VEGFA and IL-8, in both the GTL-16 and U87MG models. Effects were observed after only one to two days of PF-2341066 administration indicating that they were not due to differences in tumor burden. This data suggests that antiangiogenic activity observed with PF-2341066 may be mediated by direct or indirect mechanisms which may have implications in terms of the breadth of spectrum of antiangiogenic activity across tumor types.

In summary, these studies illustrate the pharmacologic effects of PF-2341066 on a variety of pro-oncogenic tumor cell phenotypes, tumor models, and molecular mediators of cancer progression. The broad antitumor efficacy and good safety profile of this molecule suggests its potential as a novel agent for treatment of a variety of cancers. In addition, collective results indicate that the role of c-Met in progression of cancers may differ depending on the cell lineage of origin and genetic context of a given tumor cell. The diverse role of c-Met in different tumor types as a regulator of tumor growth, survival, or metastasis will be a critical issue in the clinical development of c-Met inhibitors, monitoring of patient response, and design of clinical studies. Collectively, these results demonstrate the therapeutic potential of targeting c-Met with selective small molecule inhibitors for the treatment of human cancers.

In conclusion, cancer has traditionally been considered a disease of genetic defects so it follows that effective new treatments will selectively target the genetic defects underlying a given cancer. In general the scientific community agrees with this assessment and further agrees that only in rare instances is cancer a result of only one mutation, but more commonly is the result of accumulated molecular defects. Given the complexity of the disease, combination therapies are required and clinical success depends on proper identification of patients with genetic profiles that will exhibit the greatest predicted benefit to a selected therapeutic regimen. Inhibitors of Chk1 and c-Met have demonstrated therapeutic potential and may play significant roles in treatment of cancers in the near future.

References

1. Hartwell, L. H. and Weinert, T. A. Checkpoints: controls that ensure the order of cell cycle events. Science 1989; 246:629–634.
2. Hartwell, L H. and Kastan, M. B. Cell cycle control and cancer. Science 1994; 266: 1821–1828.
3. O'Connor, P. M. Mammalian G_1 and G_2 phase checkpoints. Cancer Surveys, 1997; 29: 151–182.
4. Sanchez, Y.; Wong, C.; Thoma, R. S.; Richman, R.; Wu, Z; Piwnica-Worms, H.; Elledge, S. J. Conservation of the Chk1 checkpoint pathway in mammals: linkage of DNA damage to Cdk regulation through Cdc25. Science 1997; 277:1497–501.
5. Zhou, B. B. and Elledge, S. The DNA damage response: putting checkpoints in perspective. Nature 2000; 408:433–9
6. Liu, Q.; Guntuku, S.; Cui, X. S.; Matsuoka, S.; Cortez, D.; Tamai, K.; Luo, G.; Carattini-Rivera, S.; DeMayo, F.; Bradley, A.; Donehower, L. A.; Elledge, S. J. Chk1 is an essential kinase that is regulated by Atr and required for the G(2)/M DNA damage checkpoint. *Genes Dev* 2000; 14:1448–59.
7. Royds, A. and Iacopetta, B. p53 and disease: when the guardian angel fails. Cell Death and Differentiation 2006; 13:1017–1026.
8. Chen, Z.; Xiao, Z.; Chen, J.; Ng, S. C.; Sowin, T. J.; Sham, H.; Rosenberg, S.; Fesik, S.; Zhang, H. Human Chk1 expression is dispensable for somatic cell death and critical for sustaining G2 DNA damage checkpoint. Mol Cancer Ther 2003; 2:543–8.
9. Zachos, G.; Rainey, M. D.; Gillespie, D. A. Chk1-deficient tumor cells are viable but exhibit multiple checkpoint and survival defects. *EMBO J* 2003; 22:713–23.
10. Zhi-Fu, T. and Nan-Horng, L. Chk1 inhibitors for novel cancer treatment. Anticancer Agents in Medicinal Chemistry, 2006; 6:377–388.
11. Braakhuis, B. J.; Ruiz van Harperen, V. W.; Boven, E., et al. Schedule-dependent antitumor effect of gemcitabine in in vivo model systems. Semin Oncol 1995; 22 (Suppl 11):42–6.
12. Houghton, P.; Stewart, C.; Zamboni, W., et al. Schedule dependent efficacy of camptothecins in models of human cancer. Ann NY Acad Sci 1996; 803:188–201.
13. Birchmeier, C.; Birchmeier, W.; Gherardi, E. and Vande Woude, G. F. Met, metastasis, motility and more. Nat Rev Mol Cell Biol 2003; 4:915–25.
14. Comoglio, P. M. and Trusolino, L. Invasive growth: from development to metastasis. J Clin Invest 2002; 109:857–62.
15. Di Renzo, M. F.; Olivero, M.; Martone, T., et al. Somatic mutations of the MET oncogene are selected during metastatic spread of human HNSC carcinomas. Oncogene 2000; 19:1547–55.

16. Ma, P. C.; Jagdeesh, S.; Jagadeeswaran, R., et al. c-MET expression/activation, functions, and mutations in non-small cell lung cancer. Proc Amer Assn Cancer Res 2004; 44:1875.
17. Ma, P. C.; Kijima, T.; Maulik, G., et al. c-MET mutational analysis in small cell lung cancer: novel juxtamembrane domain mutations regulating cytoskeletal functions. Cancer Res 2003; 63:6272–81.
18. Schmidt, L.; Duh, F. M.; Chen, F., et al. Germline and somatic mutations in the tyrosine kinase domain of the MET proto-oncogene in papillary renal carcinomas. Nature Genet 1997; 16:68–73.
19. Di Renzo, M. F.; Olivero, M.; Giacomini, A., et al. Overexpression and amplification of the met/HGF receptor gene during the progression of colorectal cancer. Clin Cancer Res 1995; 1:147–54.
20. Kuniyasu, H.; Yasui, W.; Kitadai, Y., et al. Frequent amplification of the c-Met gene in scirrhous type stomach cancer. Biochem Biophys Res Comm 1992; 189:227–32.
21. Bellusci, S.; Moens, G.; Gaudino, G., et al. Creation of an hepatocyte growth factor/scatter factor autocrine loop in carcinoma cells induces invasive properties associated with increased tumorigenicity. Oncogene 1994; 9:1091–9.
22. Jeffers, M.; Rong, S.; Anver, M., and Vande Woude, G. F. Autocrine hepatocyte growth factor/scatter factor-Met signaling induces transformation and the invasive/metastastic phenotype in C127 cells. Oncogene 1996; 13:853–6.
23. Jeffers, M.; Schmidt, L.; Nakaigawa, N., et al. Activating mutations for the met tyrosine kinase receptor in human cancer. Proc Natl Acad Sci USA 1997; 94:11445–50.
24. Rong, S. Bodescot, M.; Blair, D., et al. Tumorigenicity of the met proto-oncogene and the gene for hepatocyte growth factor. Mol Cell Biol 1992; 12:5152–8.
25. Rong, S.; Segal, S.; Anver, M.; Resau, J. H., and Vande Woude, G. F. Invasiveness and metastasis of NIH 3T3 cells induced by Met-hepatocyte growth factor/scatter factor autocrine stimulation. Proc Natl Acad Sci USA 1994; 91:4731–5.
26. Rosen, E. M. and Goldberg, I. D. Regulation of angiogenesis by scatter factor. Exs 1997; 79:193–208.
27. Rosen, E. M.; Grant, D. S.; Kleinman, H. K., et al. Scatter factor (hepatocyte growth factor) is a potent angiogenesis factor in vivo. Symposia of the Society for Experimental Biology 1993; 47:227–34.
28. Zhang, Y. W.; Su, Y.; Volpert, O. V., and Vande Woude, G. F. Hepatocyte growth factor/scatter factor mediates angiogenesis through positive VEGF and negative thrombospondin 1 regulation. Proc Natl Acad Sci USA 2003; 100:12718–23.
29. Gille, J., Khalik, M., Konig, V., and Kaufmann, R. Hepatocyte growth factor/scatter factor (HGF/SF) induces vascular permeability factor (VPF/VEGF) expression by cultured keratinocytes. J Invest Dermatol 1998; 111:1160–5.
30. Christensen, J.; Burrows, J., and Salgia, R. c-Met as a target for human cancer and characterization of inhibitors for therapeutic intervention. Cancer Letters 2005; 225:1–25.
31. Manning, G.; Whyte, D. B.; Martinez, R.; Hunter, T., and Sudarsanam, S. The protein kinase complement of the human genome. Science 2002; 298:1912–34.
32. Drexler, H. G.; Gignac, S. M.; von Wasielewski, R.; Werner, M., and Dirks, W. G. Pathobiology of NPM-ALK and variant fusion genes in anaplastic large cell lymphoma and other lymphomas. Leukemia 2000; 14:1533–59.

5
Aurora Kinases and Their Inhibitors: More Than One Target and One Drug

Patrizia Carpinelli and Jürgen Moll

Abstract Dependent on the degree of inhibition of different Aurora kinase family members, various events in mitosis are affected, resulting in differential cellular responses. These different cellular responses have to be considered in the clinical development of the small molecule inhibitors with respect to the chosen indications, schedules and appropriate endpoints. Here the properties of the most advanced small molecule Aurora kinase inhibitors are compared and a case report on the development of PHA-739358 – a spectrum selective kinases inhibitor with a dominant phenotype of Aurora kinases inhibition, which is currently being tested in clinical trials – is discussed. One of the selection criteria for this compound was its property of inhibiting more than one cancer relevant target, such as Abl wild-type and the multidrug resistant Abl T315I mutant. This opens another path for clinical development in CML, and clinical trials are underway to evaluate the activity in patients suffering from chronic myelogenous leukemia, who developed resistance to currently approved treatments.

5.1 The History of Aurora Kinases

Typically, cancer cells are characterized by an uncontrolled cell proliferation and interfering with mitosis is one of the cornerstones of cancer therapy. Current antimitotic drugs target tubulins and are highly efficacious in some cancer types, but they also target non-proliferating cells leading to side effects such as neuropathy, which is observed in patients treated with e.g. taxanes. Hence, there is a need for more specific targets interfering with mitosis (Jackson, Patrick, Dar and Huang 2007). Attractive druggable targets are kinases with essential roles in mitosis such as polo-like kinases, NEK family members or the Aurora kinases.

Nerviano Medical Sciences Srl. Viale Pasteur 10, I-20014 Nerviano (Mi), Italy

F. Colotta and A. Mantovani (eds.), *Targeted Therapies in Cancer.*
© Springer 2008

The first Aurora kinase was originally described by Glover and co-workers in 1995 in a Drosophila mutant in which the loss of function of a serine-threonine protein kinase led to a failure of the centrosomes to separate and to form a bipolar spindle (Glover, Leibowitz, McLean and Parry 1995). It was another three years before the human homologue was described by Bischoff and Zhou and a first link to cancer was established (Bischoff, Anderson, Zhu, Mossie, Ng, Souza, Schryver, Flanagan, Clairvoyant, Ginther, Chan, Novotny, Slamon and Plowman 1998; Zhou, Kuang, Zhong, Kuo, Gray, Sahin, Brinkley and Sen). Soon after, two additional Aurora kinase family members were described in mammals and the pharmaceutical industry became interested in these targets, leading to the identification of small molecule inhibitors and the first description of Hesperadin five years later (Hauf, Cole, LaTerra, Zimmer, Schnapp, Walter, Heckel, van, Rieder and Peters 2003), although this initial compound did not enter clinical trials.

One year later the first patient was treated with an Aurora kinase inhibitor (PHA-739358, July 2004) rapidly followed by several other small molecule inhibitors, which will be described later in more detail. At present several phase I & II clinical trials using small molecules are ongoing and the first results are expected in 2008. It is encouraging that it took less than 10 years from the first description of the target until the first patient was treated (Table 5-1). This period was much longer for other molecular targeted therapies on the market such as inhibitors for EGFR, Abl or VEGFR and the trend towards a faster translation of new targets into medicines is expected to accelerate further in the future.

5.2 Aurora Kinases: Functions in Different Steps in Mitosis

There are three mammalian Aurora kinases and initially the nomenclature was very heterogeneous (Fancelli and Moll 2005), however, they are now designated Aurora-A, -B and -C. Aurora-C is the least characterized and its function is somehow controversial: on the one hand Aurora-C has been described to have its major role in meiosis during spermatogenesis, which is reflected by its high abundance in testis (Tang, Lin and Tang 2006). On the other hand, an overlapping function with Aurora-B has been described in mitotic cells (Li, Sakashita, Matsuzaki, Sugimoto, Kimura, Hanaoka, Taniguchi, Furukawa and Urano 2004;

Table 5-1 Key milestones

Year	
1995	Aurora kinases identified in Drosophila: Glover et al. Cell, 81, 95
1998	Aurora kinases identified in man and link to cancer: Bischoff et al. EMBO J. 17, 3052 Zhou et al Nat. Genet. 20, 189
2003	First small inhibitor reported: Ditchfield et al. J. Cell Biol. 161, 267; Hauf et al. Cell Biol. 16, 281)
2004	First Aurora inhibitor in man: PHA-739358, July 2004
2005	Further clinical trials: Merck (VX-680/MK-0457), Astra-Zeneca (AZD-1152), Millenium Pharm. (MLN-8054)
2006	Phase II trials started (NMS, Merck), first responses in CML patients

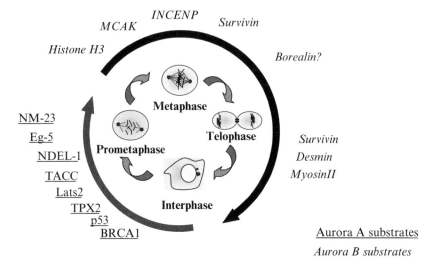

Fig. 5-1 Substrates of Aurora kinase A and B in mitosis

Sasai, Katayama, Stenoien, Fujii, Honda, Kimura, Okano, Tatsuka, Suzuki, Nigg, Earnshaw, Brinkley and Sen 2004; Tang, et al. 2006). Due to low expression levels or even lack of expression in tumors, Aurora-C is generally considered as a weakly validated target in anticancer therapy. The question of which is the right Aurora kinase to inhibit or if it is better to develop a pan-Aurora kinase inhibitor was, one of the key points, which had to be addressed early in preclinical development.

There are additional key points that any drug discovery company working in the field had to address early in preclinical development, for example, is it better to develop a pan-Aurora kinase inhibitor and which is the best route for administration (oral *vs* I.V.). Looking into the different functions of Aurora-A vs. -B (Fig. 5-1) and the cellular phenotypes when inhibiting one or the other form, it becomes clear that selectively inhibiting Aurora-A is a different approach compared to selectively inhibiting Aurora-B or Aurora-A and -B simultaneously. Elegant experiments have been performed using genetic tools and selective inhibitors (Girdler, Gascoigne, Eyers, Hartmuth, Crafter, Foote, Keen and Taylor 2006) showing that an Aurora-B selective compound phenocopies an inhibitor which acts against both Aurora-A and -B.

5.2.1 *Aurora-A*

Compared to Aurora-B, Aurora-A shows differences in sub-cellular localizations, timing of activation and functions during mitosis. Aurora-A localizes to centrosomes during interphase and moves to the spindle poles in early mitosis. Several studies in different species associated Aurora-A with a major role in centrosome maturation and spindle assembly. To exert its functions Aurora-A is associated with different proteins, many of which are substrates and many of which are altered in cancer such as BRCA1, Lats2, NM-23, p53, or TACC (reviewed in Li and Li 2006). Protein levels as well as kinase activity increase during mitosis and the peak of Aurora-A activity is seen during pro-metaphase. Moreover, the chromosomal localization of Aurora-A maps to chromosome 20q13, a locus frequently amplified in tumors. Indeed, over-expression of Aurora-A, often due to amplification of the gene, is seen in a variety of cancers.

Aurora-A has been shown to act as a weak oncogene, since over-expression of wild-type or of a constitutive active mutant of Aurora-A is able to transform Rat1 and NIH 3T3 cells as assessed by following colony formation in a soft agar assay (Bischoff, et al. 1998; Zhou, Kuang, Zhong, Kuo, Gray, Sahin, Brinkley and Sen 1998). Also NIH 3T3 cells expressing constitutively active Aurora-A can grow as solid tumors when injected into nude mice. Recently, a role for Aurora-A in stem cell self-renewal in Drosophila neuroblasts has been described, and it will be interesting to find out whether this holds true for mammalian stem cells and, in particular, for cancer stem cells (Lee, Andersen, Cabernard, Manning, Tran, Lanskey, Bashirullah and Doe 2006; Wang, Somers, Bashirullah, Heberlein, Yu and Chia 2006). Functions outside mitosis have been proposed such as activation of NF-kappaB via IkappaBalpha phosphorylation (Sun, Chan, Briassouli and Linardopoulos 2007; Briassouli, Chan, Savage, Reis-Filho and Linardopoulos 2007) or promotion of mRNA poly-adenylation (Sasayama, Marumoto, Kunitoku, Zhang, Tamaki, Kohmura, Saya and Hirota 2005) and, therefore, inhibition of Aurora-A might have additional cellular effects outside mitosis.

5.2.2 *Aurora-B*

Aurora-B is a chromosomal passenger protein which is localized to centromeres in metaphase and stays associated with the spindle midzone in anaphase. Aurora-B regulates critical steps throughout mitosis: upon entry into mitosis Aurora-B phosphorylates histone H3 on position serine^{10} followed by the control of correct biorientation and segregation of the chromosomes in metaphase, where it functions as a gatekeeper for the spindle assembly checkpoint. During prometaphase and metaphase Aurora-B associates with kinetochores where it recruits and stabilizes centromeric proteins. Aurora-B senses the lack of tension of unattached kinetochores and prevents progression from metaphase to anaphase. For the spindle checkpoint function it has been demonstrated that Aurora-B kinase activity and Bub1 cooperate to maintain the spindle checkpoint by promoting the association of BubR1 with the anaphase promoting complex/cyclosome (APC/C, Morrow, Tighe, Johnson, Scott, Ditchfield and Taylor 2005). Finally, cytokinesis cannot be completed in the absence of Aurora-B kinase activity, also indicating an essential function in completion of cell division (For review see Carmena and Earnshaw, 2003; Carvajal, Tse and Schwartz 2006). Aurora-B exerts its functions in the different phases of mitosis by phosphorylating critical components such as the chromosomal passenger complex, which consists of Aurora-B kinase, INCENP, Survivin and Borealin and which plays an important role during mitosis and cytokinesis. An essential role of the complex is to ensure that Aurora-B is accessible to phosphorylate its various substrates at the right time. Several centromeric proteins are substrates of Aurora-B including INCENP (Bishop and Schumacher, 2002), survivin (Wheatley, Henzing, Dodson, Khaled and Earnshaw 2004) and probably borealin (Gassmann, Carvalho, Henzing, Ruchaud, Hudson, Honda, Nigg, Gerloff and Earnshaw 2004). Aurora-B functions in cytokinesis might be associated with the phosphorylation of structural components necessary for cytokinesis such as vimentin (Goto, Yasui, Kawajiri, Nigg, Terada, Tatsuka, Nagata and Inagaki 2003), desmin (Kawajiri, Yasui, Goto, Tatsuka, Takahashi, Nagata and Inagaki 2003) or the myosin II regulatory light chain (Murata-Hori, Fumoto, Fukuta, Iwasaki, Kikuchi, Tatsuka and Hosoya 2000).

5.2.3 *Links to Cancer*

Aurora-A and -B are detectable and expressed at various levels in all human cancer cell lines investigated so far (Li, Zhu, Firozi, Abbruzzese, Evans, Cleary, Friess and Sen 2003; Gritsko, Coppola, Paciga, Yang, Sun, Shelley, Fiorica, Nicosia and Cheng 2003; Jeng, Peng, Lin and Hsu 2004) and their functional inactivation by genetic means (Ditchfield, Johnson, Tighe, Ellston, Haworth, Johnson, Mortlock, Keen and Taylor 2003; Marumoto, Honda, Hara, Nitta, Hirota, Kohmura and Saya 2003) or by small molecules has a strong impact on the proliferative capacity of cells (Hauf, et al. 2003; Harrington, Bebbington, Moore, Rasmussen, Jose-Adeogun, Nakayama, Graha, Demur, Hercend, Diu-Hercend, Su, Golec and Miller 2004). As mentioned above, the Aurora-A gene is amplified in many tumors, which is one, but not the only mechanism resulting in its over-expression. Tumors, or cell lines derived from tumors, with a substantial degree of Aurora-A gene amplification are breast (Li, et al, 2003; Sen, Zhou and White 1997), colon (Bischoff, et al. 1998), ovary (Gritsko, Coppola, Paciga, Yang, Sun, Shelley, Fiorica, Nicosia and Cheng 2003), gastric (Sakakura, Hagiwara, Yasuoka, Fujita, Nakanishi, Masuda, Shimomura, Nakamura, Inazawa, Abe and Yamagishi 2001) and pancreatic tumors (Li, et al. 2003). In medulloblastoma, Aurora-A over-expression is an independent predictor of patient survival (Neben, Korshunov, Benner, Wrobel, Hahn, Kokocinski, Golanov, Joos and Lichter 2004).

In cancer patients Aurora-A was reported to be a candidate low-penetrance tumor susceptibility gene. It shows a polymorphism in position 31(phenylalanine or isoleucine) and the isoleucine form was found to be preferentially expressed in some cancers, such as ovarian carcinoma (DiCioccio, Song, Waterfall, Kimura, Nagase, McGuire, Hogdall, Shah, Luben, Easton, Jacobs, Ponder, Whittemore,

Gayther, Paul Pharoah and Kruger-Kjaer 2004). The same polymorphism is associated with the advanced disease status of esophageal squamous cell carcinoma (Miao, Sun, Wang, Zhang, Tan and Lin 2004). Together with a second polymorphism (val57), an association with an increased risk in breast cancer was identified (Egan, Newcomb, Ambrosone, Trentham-Dietz, Titus-Ernstoff, Hampton, Kimura and Nagase 2004). The val57 polymorphism is also associated with an increased risk in cancer progression in gastric cancer (Ju, Cho, Kim, Kim, Ihm, Noh, Kim, Hahn, Choi and Kang 2006).

At present no polymorphism of Aurora-B has been reported in cancer, although over-expression in colorectal cancer has been described (Tatsuka, Katayama, Ota, Tanaka, Odashima, Suzuki, and Terada, 1998) and *in vitro* over-expression leads to defects in chromosome separation and *in vivo* invasiveness is increased.

Aurora-A over-expression correlates with p53 mutations in hepatocellular carcinoma and tumors having both alterations have the worst prognosis (Jeng, et al. 2004). A direct link between p53 and Aurora exists since p53 is a direct substrate of Aurora-A and is phosphorylated in position 215 (Liu, Kaneko, Yang, Feldman, Nicosia, Chen and Cheng 2004) and 315 (Katayama, Sasai, Kawai, Yuan, Bondaruk, Suzuki, Fujii, Arlinghaus, Czerniak and Sen 2004), leading to suppression of its transcriptional activity and acceleration of its MDM-2 mediated degradation, respectively.

Inhibition of Aurora-A, therefore, might lead to stabilization and transcriptional activation of p53 and expression of its downstream transcriptional targets such as p21, which inhibits CDK activity. Recently, deletion of p53 has been correlated with down-regulation or deletion of Aurora-A, which further supports the concept of a cross-talk with p53 (Mao, Wu, Perez-Losada, Jiang, Li, Neve, Gray, Cai and Balmain 2007).

5.3 Advanced Small Molecule Aurora Kinase Inhibitors

As of today, it is estimated that more than 20 programs have been initiated for the development of small molecule Aurora inhibitors, and the majority of large pharmaceutical companies active in Oncology have or have had activities running in the field. As mentioned above, one of the major questions in developing small molecule inhibitors is whether specific or dual Aurora-A or –B inhibitors will be the best choice for having the best anticancer effect. But besides the kinase inhibition profile, additional factors need to be taken into account: will an acute (I.V.) or chronic (oral) treatment be the optimal strategy taking into consideration the expected and clinically observed dose limiting toxicity, which is myelosuppression? Drug-likeness and opportunistic cross-reactivities with other cancer relevant kinases might have a strong impact on the differentiation of the multiple small molecule inhibitors, which have entered clinical trials. All these factors lead to a strong stratification of the compounds currently in development (Table 5-2) and the most advanced compounds will be discussed below in more detail.

Table 5-2 Aurora Inhibitors in Development

Compound	Chemical class	Selectivity profile	Stage	Comments	Clinical trials.gov identifier
VX-680 MK-0457 (Merck/US)	Pyrazolo-quinazoline	Pan-Aurora	Phase II	Phase I results reported at ASCO 2006, I.V. hits Flt-3, T3151 Abl & Jak-2	NCT00104351 NCT00099346 NCT00405054 (CML[1]) NCT00111683 NCT00290550 (NSCLC[2]) NCT00099346 (CRC[3])
PHA-739358 (NMS)	Pyrrolo-pyrazole	Pan-Aurora	Phase II	Multi-targeted inhibitor, I.V. hits T3151 Abl, Ret, Trk-A	NCT00335868 (CML[1])
R-763/AS-703569 (Rigel/Serono/ Aventis-Sanofi)	Not disclosed	Pan-Aurora	Phase I	Oral, hits Flt-3	NCT00391521 (solid tumors). Additional phase I in hematological malignancies
MLN-8054 (Millennium Pharm.)	Benzazepine	Aurora-A >>Aurora-B	Phase I (II?)	Phase I results reported at ASCO 2006, oral	NCT00249301
Cyc-116 (Cyclacel)	Not disclosed	Aurora A & B	Phase I in 2007	IND submitted in Dec. 2006	no data
PF-03814735 (Pfizer)	Not disclosed	Aurora A & B	Phase I	Entered Phase I in Jan. 2007	NCT00424632
AT-9283 (Astex)	Not disclosed	Pan-Aurora	Phase I/IIa	Hits T3151 Abl, Jak-2, I.V. with oral possibility	no data
AZD-1152 (Astra-Zeneca)	Pyrimidine	Aurora-B >>Aurora-A	Phase I	Phase I results reported at ASCO 2006, I.V. pro-drug	NCT00338182

[1] Chronic Myeloid Leukemia, [2] Non Small Cell Lung Cancer, [3] Colorectal Cancer

AZD-1152 is an Aurora kinase inhibitor with higher potency on Aurora-B compared to Aurora-A. The compound is a phosphate-derivative of a quinazoline and acts as a pro-drug. It is currently in Phase I in advanced solid tumors where it is administered as a two hour infusion on day 1, 8 and 15 of a 28-day cycle. Initial results showed that neutropenia is the DLT and a MTD of 200 mg has been reached with no objective responses, but some stable diseases were observed (Schellens, Boss, Witteveen, Zandvliet, Beijnen, Voogel-Fuchs, Morris, Wilson and Voest 2006).

VX-680/MK-0457 is a pyrimidine derivative and inhibits all Aurora kinase family members with cross-reactivities for Flt-3 and Abl including the Abl T315I mutant (Harrington, et al. 2004; Young, Shah, Chao, Seeliger, Milanov, Biggs III, Treiber, Patel, Zarrinkar, Lockhart, Sawyers and Kuriyan 2006). The compound has been assessed in Phase I clinical trials in patients with advanced solid tumors and was administered by a continuous five day IV infusion every four weeks. The DLT turned out to be febrile neutropenia and, also in this case, occasional stable diseases were found in patients suffering in NSCLC or pancreatic carcinoma. Phase II studies to determine the activity of the compound in patients with hematological malignancies including relapsed or refractory AML, ALL, myelodysplastic syndrome and CML have been initiated and Phase II single-agent studies in solid tumors such as NSCLC and colorectal cancer have been started. Since the compound has activity on the T315I Abl mutant, an alternative development path has been entered for CML. Forty patients have been treated by continuous five-day infusions every two to three weeks at doses of 8-32 mg/m^2 per hour in a Phase I trial including 15 patients with accelerated or blast phase CML, of which 11 carried the T315I mutation. All T315I patients responded to the therapy with one major and four minor hematological responses and one complete and two partial cytogenetic responses observed (Hampton 2007). The average treatment duration was three months with one patient treated for 15 cycles lasting for seven and one-half months. Adverse events rather resemble those observed for classical cytotoxics such as hair loss, nausea, skin rash and water retention, as well as dose-dependent bone marrow suppression. Clinical responses were also observed in a Ph$^+$ ALL patient resistant to previous treatments with Imatinib, Dasatinib or Nilotinib (Giles, Cortes, Jones, Bergstrom, Kantarjian, and Freedman 2007). In January 2007, Merck initiated a Phase II trial for CML or Ph$^+$ ALL patients with T315I Abl mutation, which failed previous treaments. The trial, which is expected to enroll 270 patients, is foreseen to support registration in the United States. Four cohorts are treated using the schedule of five days of I.V. infusion every two to three weeks. The first three cohorts are comprised of patients with accelerated phase CML, patients with blast CML and patients with chronic CML, whereas the fourth group are Ph$^+$ ALL patients (see *www.clinicaltrials.gov*, NCT00405054). Also combination trials with Dasatinib or Nilotinib are planned before the end of 2007.

Astex Therapeutics Ltd. has entered clinical phase I/IIa trials with an Aurora-A and -B kinase inhibitor (**AT-9283**) for which the chemical structure has not been disclosed to date. The compound is being developed as an injectable compound in hematological malignancies including relapsed or refractory AML, CML and myelodysplastic disorders. Since the compound shows cross-reactivity with Jak-2 and Abl, patients bearing the activated mutation of Jak-2 and CML patients with the T315I Abl mutation are expected to benefit (see *www.astex-therapeutics.com*).

MLN-8054 is an orally available inhibitor with activity mainly on Aurora-A developed by Millennium Pharmaceuticals (Manfredi, Ecsedy, Meetze, Balani, Burenkova, Chen, Galvin, Hoar, Huck, Leroy, Ray, Sells, Stringer, Stroud, Vos, Weatherhead, Wysong, Zhang, Bolen and Claiborne 2007). The compound entered clinical trial in 2005 and was investigated in phase I in solid tumors including lymphomas (Galvin, Huck, Burenkova, Burke, Bowman, Shinde, Stringer, Zhang, Manfredi and Meetze 2006). Indications followed are breast, colon, pancreas and bladder cancer (see *www.clinicaltrial.gov*, NCT00249301).

Additional compounds which have entered or are supposed to enter clinical studies have been developed by Rigel/Serono (**R763/AS-703569**), and Cyclacel (**Cyc-116**). In December 2006, Cyclacel submitted an IND to start clinical trials with their lead compound in the Aurora kinase inhibitor program, which is an orally-active inhibitor of Aurora-A and -B and VEGFR2. Pfizer entered clinical trials in 2007 with **PF-03814735** that is administered orally as a single agent in patients with advanced solid tumors (see *www.clinicaltrial.gov*, NCT00424632). A series of Aurora compounds have been described for which little information is available at present but which are expected to be in different stages of preclinical or clinical development (**SNS-314**/Sunesis Pharmaceuticals, **MP-529**, Supergen/ Montigen; **MKC-1260** and **MKC-1693**/Entremed & Miikana Therapeutics; **AB-038** BMS/Ambit; **AX-39459** GPC Biotech AG).

The wide range of companies developing Aurora kinase inhibitors is a measure of how attractive these kinases are as targets for drug discovery. But the easy drugability certainly plays a role, which might be related to the high flexibility of the protein as can be concluded from the numerous protein crystal structures solved in complex with different small molecules.

5.4 A Case History: PHA-739358

In an attempt to identify an inhibitor of Aurora kinases, in the following chapters the characteristics of the small molecule inhibitor PHA-739358 are described, which is at present in phase II clinical trials.

5.4.1 Biochemical and Cellular Activities

As part of a program at Nerviano Medical Sciences Srl. towards the development of anticancer drugs targeting kinases, we have designed new molecules based on the 3-aminopyrazole moiety, a well-known adenino mimetic pharmacophore present in several classes of kinase inhibitors. The hydrogen-bonding pattern of this class is

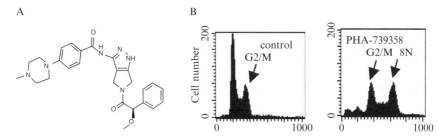

Fig. 5-2 A) Chemical structure of PHA-739358 (N-[5-(2-Methoxy-2-phenyl-acetyl)-1,4,5,6-tetrahydro-pyrrolo[3,4-c]pyrazol-3-yl]-4-(4-methyl-piperazin-1-yl)-benzamide). B) Cell cycle profile of the colorectal cancer cell line HCT-116 treated for 24 hours with 400 nM PHA-739358

well adapted to interact with the hinge region, which is highly conserved across all the kinases. The diverse elements point into specific pockets of the ATP binding site and can be used to modulate potency and selectivity towards different kinases. The crystal structures of several compounds of this class have been resolved as a complex with the kinase domain of Aurora-A (Fancelli, Moll, Varasi, Bravo, Artico, Berta, Bindi, Cameron, Candiani, Cappella, Carpinelli, Croci, Forte, Giorgini, Klapwijk, Marsiglio, Pesenti, Rocchetti, Roletto, Severino, Soncini, Storici, Tonani, Zugnoni and Vianello 2006) and they bind in the expected mode and make the expected hydrogen-bonding interactions with the hinge region. As a result of lead optimization, we selected PHA-739358 (Fig. 5-2A) for further development as an inhibitor of all Aurora kinases and with cross-reactivities relevant to cancer.

PHA-739358 shows a low nano-molar activity towards the Aurora kinase family members inhibiting Aurora-A -B and -C at 13 nM, 79 nM and 61 nM, respectively. When tested against a panel of 32 additional kinases, four kinases scored <10-fold relative to the IC50 for Aurora-A kinase; PHA-739358 shows cross-reactivity with Ret, TrkA and Abl (2-fold IC50), and FGFR1 (4-fold IC50). These cross-reactivities might increase the usability of the compound by increasing antitumor activity or extending the indications since, for example, Abl is the major driver in the majority of CML and a subset of ALL patients (Clarkson, Strife, Wisniewski, Lambek and Liu 2003; Maurer, Janssen, Thiel, van, Ludwig, Aydemir, Heinze, Fonatsch, Harbott and Reiter 1991). Expression of Ret and TrkA has been linked to thyroid cancers (Bongarzone, Vigneri, Mariani, Collini, Pilotti and Pierotti 1998) and a role for TrkA in prostate cancer has been suggested (Dionne, Camoratto, Jani, Emerson, Neff, Vaught, Murakata, Djakiew, Lamb, Bova, George and Isaacs 1998). Recently, Ret has been identified as one of the genes most frequently altered in breast cancer (Sjoblom, Jones, Wood, Parsons, Lin, Barber, Mandelker, Leary, Ptak, Silliman, Szabo, Buckhaults, Farrell, Meeh, Markowitz, Willis, Dawson, Willson, Gazdar, Hartigan, Wu, Liu, Parmigiani, Park, Bachman, Papadopoulos, Vogelstein, Kinzler and Velculescu 2006). Anti-proliferative activity of PHA-739358 has been observed in a variety of tumor cell lines in the low nano-molar to sub-micromolar range and treated cells usually show an accumulation in the number of cells with a 4N DNA content and often sub-populations of >4N are observed (Fig. 5-2B).

One explanation for the differential response might be due to the genetic background of the tumor cells tested and p53 has been shown to be a primary candidate for this effect, since cell lines defective in p53 were shown to be more prone to enter endoreduplication cycles when treated with an Aurora kinases inhibitor (Ditchfield, et al. 2003). We tested our compounds in wild-type and p53-deficient mouse embryonic fibroblasts and, whereas the wild-type cells underwent an arrest with 4N DNA content, the p53-deficient cells entered endoreduplication and accumulated with >4N DNA content. Since p53 plays an important role in G1 arrest in response to tetraploidization (Lanni and Jacks 1998), these findings suggest that the polyploidy phenotype arises through a defective cytokinesis e.g. due to inhibition of Aurora kinase B in combination with the inability of cells to prevent further cell cycle progression because of a lack or inactivating mutations of p53 or a defect in other components of the p53 pathway. A direct link between p53 and Aurora-A kinase has been reported since p53 is a direct substrate for Aurora-A and p53 is post-translationally stabilized after Aurora-A down-regulation (Katayama, et al. 2004; Liu, et. al. 2004). In agreement with these findings PHA-739358 increases p53 protein levels associated with an increase in levels of p21, which is known to be regulated by p53 at the transcriptional level (el-Deiry, Harper, O'Connor, Velculescu, Canman, Jackman, Pietenpol, Burrell, Hill, Wang, Wiman, Mercer, Kastan, Kohn, Elledge, Kinzler and Vogelstein 1994).

Inhibition of Aurora-A or -B can be followed at the cellular level by determination of Aurora-A auto-phosphorylation or by inhibition of phosphorylation of histone H3. Histone H3, a protein implicated in chromosome condensation, is phosphorylated at serine^{10} by Aurora-B during mitosis (Crosio, Fimia, Loury, Kimura, Okano, Zhou, Sen, Allis and Sassone-Corsi 2002). Treatment of Hela cells with PHA-739358 led to a decrease in histone H3 phosphorylation and Aurora-A auto-phosphorylation that correlates well with the IC50s for Aurora-A and -B in biochemical assays, suggesting a potent inhibition of both kinases at nano-molar concentrations in cells.

The Ret kinase inhibitory activity of PHA-739358 was evaluated in cells which contain a Ret allele with a constitutively activating mutation in the extracellular domain (Carlomagno, Salvatore, Santoro, de Franciscis, V, Quadro, Panariello, Colantuoni and Fusco 1995) and which can be used to determine receptor auto-phosphorylation. Trk-A kinase inhibitory activity was evaluated in cells that allow the inhibition of NGF-dependent Trk-A phosphorylation to be determined. Both kinases were inhibited at low micro-molar concentrations, although sensitivity was lower compared to Aurora inhibition. PHA-739358 has an effect on MAPK activation induced by FGF, but not by EGF, demonstrating selectivity for inhibition of the FGFR1 pathway, but not the EGFR pathway.

The activity on wild-type and mutant ABL has been investigated in more detail. PHA-739358 binds wild-type Abl with high affinity as well as is binding to several mutants, including the T315I mutant for which at present no treatment option exists if allogenic bone marrow transplantation is excluded. This mutation is predicted to play a critical role in the future, since there will be a growing medical need for new treatments for this group of patients. Indeed, in CML patients increased resistance to treatments is observed at a rate of 4 percent per treatment year (Kling 2006).

As seen in the crystal structure of Abl (T315I) in complex with PHA-739358, the compound associates with the catalytic domain of an active and phosphorylated conformation of Abl (T315I) and lacks the steric hindrance imposed by the substitution of threonine by isoleucine which is, for example, seen with Imatinib (Gleevec, Glivec, STI571). Although second generation inhibitors have recently been registered for the treatment of patients resistant to Imatinib (Dasatinib, Nilotinib), none of them inhibits the Abl (T315I) mutant (Kantarjian, Giles, Wunderle, Bhalla, O'Brien, Wassmann, Tanaka, Manley, Rae, Mietlowski, Bochinski, Hochhaus, Griffin, Hoelzer, Albitar, Dugan, Cortes, Alland and Ottmann 2006; Talpaz, Shah, Kantarjian, Donato, Nicoll, Paquette, Cortes, O'Brien, Nicaise, Bleickardt, Blackwood-Chirchir, Iyer, Chen, Huang, Decillis and Sawyers 2006). The potent inhibition of Abl kinase activity seen in biochemical and cellular assays is also seen in primary CD34+ tumor cells taken from CML patients bearing the Abl T315I mutation, and promising clinical responses were also observed in patients. Other Aurora kinases inhibitors such as AT-9283 and VX-680 (MK-0457) also cross-react with Abl, and VX-680 (MK-0457) is active ex-vivo against cells from patients bearing the Abl T315I mutations (Young, et al. 2006) and clinical activity in this subset of CML patients has been reported (Giles, et al. 2007).

5.4.2 Antitumor Activity In Vivo

The antitumor activity of PHA-739358 *in vivo* was evaluated in several solid human tumor xenograft models in nude mice and a significant activity was observed. Efficacy experiments were performed in models including syngenic, transgenic and carcinogen-induced tumor models in different species to monitor species-related effects on efficacy or toxicity. Usually the compound was administered intravenously twice a day at the maximally efficacious dose of 30 mg/kg for dosing periods ranging from five to ten days. A dose-finding study was performed in HL60 leukemia cells implanted sub-cutaneously in SCID mice. After intravenous administration, at doses between 7.5 and 30 mg/kg I.V. BID for five days, a significant tumor growth inhibition was observed showing dose-dependency and a tumor growth inhibition (TGI) of up to 98 percent at the highest dose. There was also evidence of tumor regression and occasional cures (Fancelli, et al. 2006). Having observed similar activity on ovarian and colon tumor xenograft, we also tested PHA-739358 activity on two breast tumor models represented by a DMBA carcinogen-induced mammary carcinoma in rats and an activated Ras-driven mammary tumors in transgenic mice (MMTV-v-Ha-Ras). In both models, which have the advantage of bearing many similarities with human breast cancer, good activity in terms of TGI was observed demonstrating the effectiveness of PHA-739358 against syngenic models of mammary cancers arising from either oncogene activation or genetic insult by carcinogen exposure in rodents. The transgenic mouse prostate (TRAMP) model was used to evaluate the efficacy of PHA-739358 against a prostate cancer model (Greenberg, DeMayo, Finegold, Medina, Tilley, Aspinall, Cunha,

Donjacour, Matusik and Rosen 1995). After detection of tumor onset, which was monitored by MRI (Magnetic Resonance Imaging), mice were treated for five days BID with 30 mg/kg of the compound. As in the human disease, prostate tumors in TRAMP mice manifest as islands of low MRI signal intensity surrounded by high MRI signals from neighboring benign tissue, and after detection the tumor grows very fast and doubles its volume in about five days. The treatment-induced tumor regression in some animals with long-term stabilizations was observed in the majority of treated animals.

5.4.3 In Vivo Biomarker Modulation

Changes in protein expression, phosphorylation of different markers and cellular morphology were tested in different cell lines *in vitro* and *in vivo*. Morphological changes are exemplified in U2OS cells (Fig. 5-3, left panel), and in xenograft colorectal tumors (HCT-116) treated for five days with PHA-680632, (a close homologue of PHA-739358) at a dose of 30 mg/kg, BID (Fig. 5-3, right panel). The last

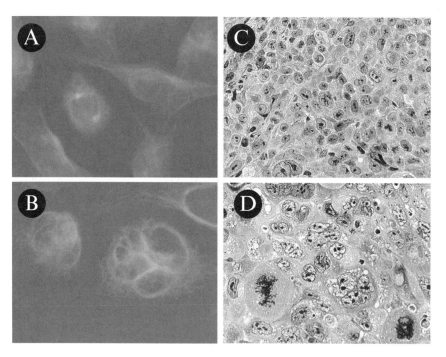

Fig. 5-3 Cellular effects of treatment with an Aurora kinase inhibitor. Induction of endoreduplication in U2OS cells treated *in vitro* with PHA-680632 (B), untreated cells are shown in A. Tissue section of HCT-116 tumors derived from xenografts in mice without (C) or after treatment with the compound (D) for 5 days

injection of the compound was performed two hours before the animals were sacrificed. Histological evaluation of treated tumor cells in both settings revealed an increase in cellular and nuclear size and the presence of multinucleated cells, similar to what is seen with PHA-739358.

Immunohistochemistry staining was performed for phospho-histone H3, p53 and p21 to demonstrate target modulation, BrdU incorporation and cyclin A levels were measured to follow cell proliferation. Caspase-3 was used to follow apoptosis. The treated tumors showed a decrease in both the number and the intensity of phospho-histone H3 positive cells, that is in agreement with what we observed by Western blot analysis of tumors from treated compared to untreated mice. The observed strong decrease in histone H3 phosphorylation by PHA-739358 represents a potential biomarker that could be useful for detecting inhibition of Aurora-B kinase activity in the clinic. BrdU incorporation and the number of cyclin A positive cells were also decreased. There was a slight increase in active caspase-3 positive cells, scattered in viable areas, in treated tumors. Interestingly, the number of macrophages that may be attracted by apoptotic cells was also increased in treated tumors, and p53 and p21 were strongly up-regulated by the compound.

The modulation of histone H3 phosphorylation by PHA-739358 has a strong potential as a biomarker since the epitope is extremely abundant in mitotic cells and with the availability of a highly sensitive and selective antibody means that very few cycling cells are necessary for detection of the signal in Western blot analysis. As shown in Fig. 5-4, we demonstrated this *in vivo* modulation not only in tumors but also in bone marrow and skin of PHA-739358 treated mice. Thirty minutes after a single injection by I.V. bolus of PHA-739358 (45 mg/kg), a reduction in Serine^{10} phosphorylation of histone H3 in A2780 tumor xenografts, was seen with full inhibition at two hours, followed by recovery at eight hours. The effect of PHA-739358 on histone H3 in skin was less pronounced relative to the effects seen in bone marrow or human tumor cell lines implanted in nude mice. Nevertheless, based on these results, skin punch biopsies before and after treatment of patients were judged feasible to semi-quantitatively assess target inhibition by PHA-739358 in the clinic. Skin biopsies were taken from patients enrolled in the two Phase I clinical trials and paired pre- and post-infusion samples were analyzed for phosphorylation of histone

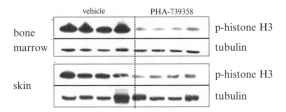

Fig. 5-4 Western blots of tissue from mice treated with PHA-739358. Modulation of histone H3 phosphorylation at serine^{10} in mice treated with 45 mg/kg of PHA-739358 given as I.V. bolus. Tissues were taken two hours after injection. Western blot of bone marrow and skin of four mice injected with vehicle or compound. Tubulin signals were used for normalization

H3 in position serine[10]. A decrease in the phosphorylation of histone H3 was observed starting from the dose levels of 190 mg/m2 after 6h infusion and 500 mg/m2 after 24h infusion (see www.clinicaltrials.gov). In these patients the steady state plasma concentration of the compound during the infusion reached the predicted plasma concentrations as calculated using a preclinical PK/PD model (Simeoni, Magni, Cammia, De, Croci, Pesenti, Germani, Poggesi and Rocchetti 2004).

5.4.4 Pharmacokinetics

The pharmacokinetics properties of PHA-739358 have been investigated in the mouse, rat, dog and monkey. In all species the compound showed moderate to high systemic clearance and a high volume of distribution, suggesting extensive tissue distribution. Following both I.V. bolus and infusion administration, PHA-739358 plasma levels increased largely in proportion to dose and data from repeated dose studies indicating that the pharmacokinetics of PHA-739358 were not time-dependent and did not show important differences when different schedules of treatment (i.e., duration of infusion) were applied. Single dose I.V. administration of PHA-739358 to mice resulted in a time-dependent inhibition of histone H3 phosphorylation measured in bone marrow. Strong inhibition at one hour post-administration was still apparent at three hours and returned to baseline levels by eight hours.

Similar observations for the inhibition of histone H3 phosphorylation were obtained after single I.V. administration of the compound both in tumor and skin tissue lysates. A PK/PD model relating the time-concentration profiles of PHA-739358 with the tumor growth was developed for A2780 xenografts. In this model it is assumed that the cells, after being exposed to the drug, are damaged and the severity of the damage (potency) and the rate at which the cells die (cell death distribution) are the critical parameters of the model and are calculated after fitting to experimental data. The model successfully predicted the response of tumors exposed to different treatment schedules with PHA-739358. Based on this PK/PD approach, the efficacy of PHA-739358 was essentially found to be related to the total amount of the compound administered in the exponential phase of the tumor growth.

5.5 Conclusion

The existence of more than 20 Aurora inhibitors in different stages of development leads to the question of whether there is space for all of these compounds. Keeping in mind the attrition rates in oncology, which are high with an estimated five percent of drugs entering clinical trials actually reaching the market, most of these compounds will not survive clinical development, and the remaining ones will try to find their niches. Stratifications could be based on (i) the selectivity profile for

Aurora kinase A and B. Certainly drugs inhibiting Aurora-B or all Aurora kinases have a different mechanism of action compared to selective Aurora-A inhibitors, (ii) the cross-reactivity profile apart from the Aurora kinases could make a difference, for example as the exciting opportunity in CML patients bearing the T315I Abl mutation and (iii) the administration route oral vs. I.V. although as of today the toxicity profile seems not to favor prolonged oral treatments. Finally, not to forget the drug-likeness of the compounds with corresponding differences in ADME properties, PK or MOA unrelated toxicities and, last but not least, the appropriate clinical development strategy and the identification of responsive patient populations or indications. The final proof of concept in solid tumors for inhibition of Aurora kinases as a potent anticancer strategy still needs to be proven but the numerous ongoing clinical trials will give this answer in the very near future.

"Personal experience, intuition, serendipity and luck still play a role alongside rigorous science and rational thinking" has been stated recently (Drews 2006), and are the critical factors for successful drug development (besides money), even when there might be different opinions on the relative weight and relevance of the single aspects.

Acknowledgments We thank the Aurora project team and, in particular, M. Rocchetti, M. L. Giorgini, P. Cappella, L. Gianellini and R. Ceruti for their contributions, and S. Healy for critical reading of the manuscript.

References

Bischoff, J. R.; Anderson, L.; Zhu, Y.; Mossie, K.; Ng, L.; Souza, B.; Schryver, B.; Flanagan, P.; Clairvoyant, F.; Ginther, C.; Chan, C. S.; Novotny, M.; Slamon, D. J., and Plowman, G. D. (1998) A homologue of Drosophila aurora kinase is oncogenic and amplified in human colorectal cancers. EMBO J. 17, 3052–3065.

Bishop, J. D. and Schumacher, J. M. (2002) Phosphorylation of the carboxyl terminus of inner centromere protein (INCENP) by the Aurora B Kinase stimulates Aurora B kinase activity. J. Biol. Chem. 277, 27577–27580.

Bongarzone, I.; Vigneri, P.; Mariani, L.; Collini, P.; Pilotti, S., and Pierotti, M. A. (1998) RET/NTRK1 rearrangements in thyroid gland tumors of the papillary carcinoma family: correlation with clinicopathological features. Clin. Cancer Res. 4, 223–228.

Briassouli, P.; Chan, F.; Savage, K.; Reis-Filho, J. S., and Linardopoulos S. (2007) Aurora-A regulation of nuclear factor-kappaB signaling by phosphorylation of IkappaBalpha. Cancer Res. 67, 1689–95.

Carlomagno, F.; Salvatore, D.; Santoro, M.; de, Franciscis, V.; Quadro, L.; Panariello, L.; Colantuoni, V., and Fusco, A. (1995) Point mutation of the RET proto-oncogene in the TT human medullary thyroid carcinoma cell line. Biochem. Biophys. Res. Commun. 207, 1022–1028.

Carmena, M. and Earnshaw, W. C. (2003) The cellular geography of aurora kinases. Nat. Rev. Mol. Cell Biol. 4, 842–854.

Carvajal, R. D.; Tse, A., and Schwartz, G. K. (2006) Aurora kinases: new targets for cancer therapy. Clin. Cancer Res. 12, 6869–6875.

Clarkson, B.; Strife, A.; Wisniewski, D.; Lambek, C. L., and Liu, C. (2003) Chronic myelogenous leukemia as a paradigm of early cancer and possible curative strategies. Leukemia 17, 1211–1262.

Crosio, C.; Fimia, G. M.; Loury, R.; Kimura, M.; Okano, Y.; Zhou, H.; Sen, S.; Allis, C. D., and Sassone-Corsi, P. (2002) Mitotic phosphorylation of histone H3: spatio-temporal regulation by mammalian Aurora kinases. Mol. Cell Biol. 22, 874–885.

DiCioccio, R. A.; Song, H.; Waterfall, C.; Kimura, M. T.; Nagase, H.; McGuire, V.; Hogdall, E.; Shah, M. N.; Luben, R. N.; Easton, D. F.; Jacobs, I. J.; Ponder, B. A. J.; Whittemore, A. S.; Gayther, S. A.; Pharoah, P. D. P., and Kruger-Kjaer, S. (2004) STK15 Polymorphisms and Association with Risk of Invasive Ovarian Cancer. Cancer Epidemiology Biomarkers & Prevention 13, 1589–1594.

Dionne, C. A.; Camoratto, A. M.; Jani, J. P.; Emerson, E.; Neff, N.; Vaught, J. L.; Murakata, C.; Djakiew, D.; Lamb, J.; Bova, S.; George, D., and Isaacs, J. T. (1998) Cell cycle-independent death of prostate adenocarcinoma is induced by the trk tyrosine kinase inhibitor CEP-751 (KT6587). Clin. Cancer Res. 4, 1887–1898.

Ditchfield, C.; Johnson, V. L.; Tighe, A.; Ellston, R.; Haworth, C.; Johnson, T.; Mortlock, A.; Keen, N., and Taylor, S. S. (2003) Aurora-B couples chromosome alignment with anaphase by targeting BubR1, Mad2, and Cenp-E to kinetochores. J. Cell Biol. 161, 267–280.

Drews, J. (2006) Case histories, magic bullets and the state of drug discovery. Nat. Rev. Drug Discov. 5, 635–640.

Egan, K. M.; Newcomb, P. A.; Ambrosone, C. B.; Trentham-Dietz, A.; Titus-Ernstoff, L.; Hampton, J. M.; Kimura, M. T., and Nagase, H. (2004) STK15 polymorphism and breast cancer risk in a population-based study. Carcinogenesis 25, 2149–2153.

el-Deiry, W. S.; Harper, J. W.; O'Connor, P. M.; Velculescu, V. E.; Canman, C. E.; Jackman, J.; Pietenpol, J. A.; Burrell, M.; Hill, D. E.; Wang, Y.;Widman, K.;G. Mercer, W.;E. Kastan, M. B.; Kohn, K. W.; Elledge, S. J.; Kinzler, K. W., and Vogelstein, B. (1994) WAF1/CIP1 is induced in p53-mediated G1 arrest and apoptosis. Cancer Res. 54, 1169–1174.

Fancelli, D. and Moll, J. (2005) Inhibitors of Aurora kinases for the treatment of cancer. Expert Opinion on Therapeutic Patents 15, 1169–1182.

Fancelli, D.; Moll, J.; Varasi, M.; Bravo, R.; Artico, R.; Berta, D.; Bindi, S.; Cameron, A.; Candiani, I.; Cappella, P.; Carpinelli, P.; Croci, W.; Forte, B.; Giorgini, M. L.; Klapwijk, J.; Marsiglio, A.; Pesenti, E.; Rocchetti, M.; Roletto, F.; Severino, D.; Soncini, C.; Storici, P.; Tonani, R.; Zugnoni, P., and Vianello, P. (2006) 1,4,5,6-tetrahydropyrrolo-[3,4c]pyrazoles: identification of a potent Aurora kinase inhibitor with a favorable antitumor kinase inhibition profile. J. Med. Chem. 49, 7247–7251.

Galvin, K. M.; Huck, J.; Burenkova, O.; Burke, K.; Bowman, D.; Shinde, V.; Stringer, B.; Zhang, M.; Manfredi, M., and Meetze, K. (2006) Preclinical pharmacodynamic studies of Aurora-A inhibition by MLN8054. Journal of Clinical Oncology, ASCO Annual Meeting Proceedings (June 20 Supplement), 24, 13059.

Gassmann, R.; Carvalho, A.; Henzing, A. J.; Ruchaud, S.; Hudson, D. F.; Honda, R.; Nigg, E. A.; Gerloff, D. L., and Earnshaw, W. C. (2004) Borealin: a novel chromosomal passenger required for stability of the bipolar mitotic spindle. J. Cell Biol. 166, 179–191.

Giles, F. J.; Cortes, J.; Jones, D.; Bergstrom, D.; Kantarjian, H., and Freedman, S. J. (2007) MK-0457, a novel kinase inhibitor, is active in patients with chronic myeloid leukemia or acute lymphocytic leukemia with the T315I BCR-ABL mutation. Blood 109, 500–502.

Girdler, F.; Gascoigne, K. E.; Eyers, P. A.; Hartmuth, S.; Crafter, C.; Foote, K. M.; Keen, N. J., and Taylor, S. S. (2006) Validating Aurora-B as an anticancer drug target. J. Cell Sci. 119, 3664–3675.

Glover, D. M.; Leibowitz, M. H.; McLean, D. A., and Parry, H. (1995) Mutations in Aurora prevent centrosome separation leading to the formation of monopolar spindles. Cell 81, 95–105.

Goto, H.; Yasui, Y.; Kawajiri, A.; Nigg, E. A.; Terada, Y.; Tatsuka, M.; Nagata, K., and Inagaki, M. (2003) Aurora-B regulates the cleavage furrow-specific vimentin phosphorylation in the cytokinetic process. J. Biol. Chem. 278, 8526–8530.

Greenberg, N. M.; DeMayo, F.; Finegold, M. J.; Medina, D.; Tilley, W. D.; Aspinall, J. O.; Cunha, G. R.; Donjacour, A. A.; Matusik, R. J., and Rosen, J. M. (1995) Prostate cancer in a transgenic mouse. Proc. Natl. Acad. Sci. U.S.A 92, 3439–3443.

Gritsko, T. M.; Coppola, D.; Paciga, J. E.; Yang, L.; Sun, M.; Shelley, S. A.; Fiorica, J. V.; Nicosia, S. V., and Cheng, J. Q. (2003) Activation and over-expression of centrosome kinase BTAK/Aurora-A in human ovarian cancer. Clin. Cancer Res. 9, 1420–1426.

Hampton, T. (2007) New blood cancer therapies under study. JAMA 297, 457–458.
Harrington, E. A.; Bebbington, D.; Moore, J.; Rasmussen, R. K.; Jose-Adeogun, A. O.; Nakayama, T.; Graham, J. A.; Demur, C.; Hercend, T.; Diu-Hercend, A.; Su, M.; Golec, J. M., and Miller, K. M. (2004) VX-680, a potent and selective small-molecule inhibitor of the Aurora kinases, suppresses tumor growth in vivo. Nat. Med. 10, 262–267.
Hauf, S.; Cole, R. W.; LaTerra, S.; Zimmer, C.; Schnapp, G.; Walter, R.; Heckel, A.; van, Meel J.; Rieder, C. L., and Peters, J. M. (2003) The small molecule Hesperadin reveals a role for Aurora-B in correcting kinetochore-microtubule attachment and in maintaining the spindle assembly checkpoint. J. Cell Biol. 161, 281–294.
Jackson, J. R.; Patrick, D. R.; Dar, M. M., and Huang, P. S. (2007) Targeted anti-mitotic therapies: can we improve on tubulin agents? Nat. Rev. Cancer 7, 107–117.
Jeng, Y. M.; Peng, S. Y.; Lin, C. Y., and Hsu, H. C. (2004) Over-expression and amplification of Aurora-A in hepatocellular carcinoma. Clin. Cancer Res. 10, 2065–2071.
Ju, H.; Cho, H.; Kim, Y. S.; Kim, W. H.; Ihm, C.; Noh, S. M.; Kim, J. B.; Hahn, D. S.; Choi, B. Y., and Kang, C. (2006) Functional polymorphism 57Val>Ile of aurora kinase A associated with increased risk of gastric cancer progression. Cancer Lett. 242, 273–279.
Kantarjian, H.; Giles, F.; Wunderle, L.; Bhalla, K.; O'Brien, S.; Wassmann, B.; Tanaka, C.; Manley, P.; Rae, P.; Mietlowski, W.; Bochinski, K.; Hochhaus, A.; Griffin, J. D.; Hoelzer, D.; Albitar, M.; Dugan, M.; Cortes, J.; Alland, L., and Ottmann, O. G. (2006) Nilotinib in imatinib-resistant CML and Philadelphia chromosome-positive ALL. N. Engl. J. Med. 354, 2542–2551.
Katayama, H.; Sasai, K.; Kawai, H.; Yuan, Z. M.; Bondaruk, J.; Suzuki, F.; Fujii, S.; Arlinghaus, R. B.; Czerniak, B. A., and Sen, S. (2004) Phosphorylation by aurora kinase A induces Mdm2-mediated destabilization and inhibition of p53. Nat. Genet. 36, 55–62.
Kawajiri, A.; Yasui, Y.; Goto, H.; Tatsuka, M.; Takahashi, M.; Nagata, K., and Inagaki, M. (2003) Functional significance of the specific sites phosphorylated in desmin at cleavage furrow: Aurora-B may phosphorylate and regulate type III intermediate filaments during cytokinesis coordinatedly with Rho-kinase. Mol. Biol. Cell 14, 1489–1500.
Kling, J. (2006) Moving diagnostics from the bench to the bedside. Nat. Biotechnol. 24, 891–893.
Lanni, J. S. and Jacks, T. (1998) Characterization of the p53-dependent postmitotic checkpoint following spindle disruption. Mol. Cell Biol. 18, 1055–1064.
Lee, C. Y.; Andersen, R. O.; Cabernard, C.; Manning, L.; Tran, K. D.; Lanskey, M. J.; Bashirullah, A., and Doe, C. Q. (2006) Drosophila Aurora-A kinase inhibits neuroblast self-renewal by regulating aPKC/Numb cortical polarity and spindle orientation. Genes Dev. 20, 3464–3474.
Li, D.; Zhu, J.; Firozi, P. F.; Abbruzzese, J. L.; Evans, D. B.; Cleary, K.; Friess, H., and Sen, S. (2003) Over-expression of oncogenic STK15/BTAK/Aurora A kinase in human pancreatic cancer. Clin. Cancer Res. 9, 991–997.
Li, J. J. and Li, S. A. (2006) Mitotic kinases: the key to duplication, segregation, and cytokinesis errors, chromosomal instability, and oncogenesis. Pharmacol. Ther. 111, 974–984.
Li, X.; Sakashita, G.; Matsuzaki, H.; Sugimoto, K.; Kimura, K.; Hanaoka, F.; Taniguchi, H.; Furukawa, K., and Urano, T. (2004) Direct association with inner centromere protein (INCENP) activates the novel chromosomal passenger protein, Aurora-C. J. Biol. Chem. 279, 47201–47211.
Liu, Q.; Kaneko, S.; Yang, L.; Feldman, R. I.; Nicosia, S. V.; Chen, J., and Cheng, J. Q. (2004) Aurora-A abrogation of p53 DNA binding and transactivation activity by phosphorylation of serine 215. J. Biol. Chem. 279, 52175–52182.
Manfredi, M. G.; Ecsedy, J. A.; Meetze, K. A.; Balani, S. K. Burenkova, O.; Chen, W.; Galvin, K. M.; Hoar, K. M.; Huck, J. J.; Leroy, P. J.; Ray, E. T.; Sells, T. B.; Stringer, B.; Stroud, S. G.; Vos T. J.; Weatherhead, G. S.; Wysong, D. R.; Zhang, M.; Bolen, J. B., and Claiborne, C. F. (2007) Antitumor activity of MLN8054, an orally active small-molecule inhibitor of Aurora-A kinase. Proc Natl Acad Sci U S A. 104, 4106–4111.
Mao, J. H.; Wu, D.; Perez-Losada, J.; Jiang, T.; Li, Q.; Neve, R. M.; Gray, J. W.; Cai, W. W., and Balmain, A. (2007) Crosstalk between Aurora-A and p53: Frequent Deletion or Downregulation of Aurora-A in Tumors from p53 Null Mice. Cancer Cell 11, 161–173.

Marumoto, T.; Honda, S.; Hara, T.; Nitta, M.; Hirota, T.; Kohmura, E., and Saya, H. (2003) Aurora-A kinase maintains the fidelity of early and late mitotic events in HeLa cells. J. Biol. Chem. 278, 51786–51795.

Maurer, J.; Janssen, J. W.; Thiel, E.; van, Denderen J.; Ludwig, W. D.; Aydemir, U.; Heinze, B.; Fonatsch, C.; Harbott, J., and Reiter, A. (1991) Detection of chimeric BCR- ABL genes in acute lymphoblastic leukaemia by the polymerase chain reaction. Lancet 337, 1055–1058.

Miao, X.; Sun, T.; Wang, Y.; Zhang, X.; Tan, W., and Lin, D. (2004) Functional STK15 Phe31Ile polymorphism is associated with the occurrence and advanced disease status of esophageal squamous cell carcinoma. Cancer Res. 64, 2680–2683.

Morrow, C. J.; Tighe, A.; Johnson, V. L.; Scott, M. I.; Ditchfield, C., and Taylor, S. S. (2005) Bub1 and Aurora-B cooperate to maintain BubR1-mediated inhibition of APC/CCdc20. J. Cell Sci. 118, 3639–3652.

Murata-Hori, M.; Fumoto, K.; Fukuta, Y.; Iwasaki, T.; Kikuchi, A.; Tatsuka, M., and Hosoya, H. (2000) Myosin II regulatory light chain as a novel substrate for AIM-1, an aurora/Ipl1p-related kinase from rat. J. Biochem.(Tokyo) 128, 903–907.

Neben, K.; Korshunov, A.; Benner, A.; Wrobel, G.; Hahn, M.; Kokocinski, F.; Golanov, A.; Joos, S., and Lichter, P. (2004) Microarray-based screening for molecular markers in medulloblastoma revealed STK15 as independent predictor for survival. Cancer Res. 64, 3103–3111.

Sakakura, C.; Hagiwara, A.; Yasuoka, R.; Fujita, Y.; Nakanishi, M.; Masuda, K.; Shimomura, K.; Nakamura, Y.; Inazawa, J.; Abe, T., and Yamagishi, H. (2001) Tumor-amplified kinase BTAK is amplified and over-expressed in gastric cancers with possible involvement in aneuploid formation. Br. J. Cancer 84, 824–831.

Sasai, K.; Katayama, H.; Stenoien, D. L.; Fujii, S.; Honda, R.; Kimura, M.; Okano, Y.; Tatsuka, M.; Suzuki, F.; Nigg, E. A.; Earnshaw, W. C.; Brinkley, W. R., and Sen, S. (2004) Aurora-C kinase is a novel chromosomal passenger protein that can complement Aurora-B kinase function in mitotic cells. Cell Motil. Cytoskeleton 59, 249–263.

Sasayama, T.; Marumoto, T.; Kunitoku, N.; Zhang, D.; Tamaki, N.; Kohmura, E.; Saya, H., and Hirota, T. (2005) Over-expression of Aurora-A targets cytoplasmic polyadenylation element binding protein and promotes mRNA polyadenylation of Cdk1 and cyclin B1. Genes Cells 10, 627–638.

Schellens, J. H.; Boss, D.; Witteveen, P. O. Zandvliet, A.; Beijnen, J. H. Voogel-Fuchs, M. Morris, C. Wilson, D., and Voest, E. E. (2006) Phase I and pharmacological study of the novel Aurora kinase inhibitor AZD1152. Journal of Clinical Oncology, ASCO Annual Meeting Proceedings (June 20 Supplement), 24, 3008.

Sen, S.; Zhou, H., and White, R. A. (1997) A putative serine/threonine kinase encoding gene BTAK on chromosome 20q13 is amplified and over-expressed in human breast cancer cell lines. Oncogene 14, 2195–2200.

Simeoni, M.; Magni, P.; Cammia, C.; De, Nicolao G.; Croci, V.; Pesenti, E.; Germani, M.; Poggesi, I., and Rocchetti, M. (2004) Predictive pharmacokinetic-pharmacodynamic modeling of tumor growth kinetics in xenograft models after administration of anticancer agents. Cancer Res. 64, 1094–1101.

Sjoblom, T.; Jones, S.; Wood, L. D.; Parsons, D. W.; Lin, J.; Barber, T. D.; Mandelker, D.; Leary, R. J.; Ptak, J.; Silliman, N.; Szabo, S.; Buckhaults, P.; Farrell, C.; Meeh, P.; Markowitz, S. D.; Willis, J.; Dawson, D.; Willson, J. K.; Gazdar, A. F.; Hartigan, J.; Wu, L.; Liu, C.; Parmigiani, G.; Park, B. H.; Bachman, K. E.; Papadopoulos, N.; Vogelstein, B.; Kinzler, K. W., and Velculescu, V. E. (2006) The consensus coding sequences of human breast and colorectal cancers. Science 314, 268–274.

Sun, C.; Chan, F.; Briassouli, P., and Linardopoulos, S. (2007) Aurora kinase inhibition downregulates NF-kappaB and sensitizes tumor cells to chemotherapeutic agents. Biochem. Biophys. Res. Commun. 352, 220–225.

Talpaz, M.; Shah, N. P.; Kantarjian, H.; Donato, N.; Nicoll, J.; Paquette, R.; Cortes, J.; O'Brien, S.; Nicaise, C.; Bleickardt, E.; Blackwood-Chirchir, M. A.; Iyer, V.; Chen, T. T.; Huang, F.; Decillis, A. P., and Sawyers, C. L. (2006) Dasatinib in imatinib-resistant Philadelphia chromosome-positive leukemias. N. Engl. J. Med. 354, 2531–2541.

Tang, C. J.; Lin, C. Y., and Tang, T. K. (2006) Dynamic localization and functional implications of Aurora-C kinase during male mouse meiosis. Dev. Biol. 290, 398–410.

Tatsuka, M.; Katayama, H.; Ota, T.; Tanaka, T.; Odashima, S.; Suzuki, F., and Terada, Y. (1998) Multinuclearity and increased ploidy caused by over-expression of the Aurora- and Ipl1-like midbody-associated protein mitotic kinase in human cancer cells. Cancer Res. 58, 4811–4816.

Wang, H.; Somers, G. W.; Bashirullah, A.; Heberlein, U.; Yu, F.; and Chia, W. (2006) Aurora-A acts as a tumor suppressor and regulates self-renewal of Drosophila neuroblasts. Genes Dev. 20, 3453–3463.

Wheatley, S. P.; Henzing, A. J.; Dodson, H.; Khaled, W., and Earnshaw, W. C. (2004) Aurora-B phosphorylation in vitro identifies a residue of survivin that is essential for its localization and binding to inner centromere protein (INCENP) in vivo. J. Biol. Chem. 279, 5655–5660.

Young, M. A.; Shah, N. P.; Chao, L. H.; Seeliger, M.; Milanov, Z. V.; Biggs, W. H., III; Treiber, D. K.; Patel, H. K.; Zarrinkar, P. P.; Lockhart, D. J.; Sawyers, C. L., and Kuriyan, J. (2006) Structure of the kinase domain of an imatinib-resistant Abl mutant in complex with the Aurora kinase inhibitor VX-680. Cancer Res. 66, 1007–1014.

Zhou, H.; Kuang, J.; Zhong, L.; Kuo, W. L.; Gray, J. W.; Sahin, A.; Brinkley, B. R., and Sen, S. (1998) Tumor amplified kinase STK15/BTAK induces centrosome amplification, aneuploidy and transformation. Nat. Genet. 20, 189–193.

# 6
Signalling Pathways and Adhesion Molecules as Targets for Antiangiogenesis Therapy in Tumors

Gianfranco Bazzoni

6.1 Introduction

In the embryo, blood vessels derive from endothelial precursors in a process called vasculogenesis. These progenitors assemble into a primitive vascular plexus. Subsequently, in a process called angiogenesis, the primitive vascular plexus expands by means of vessel sprouting and organizes into a network of blood vessels. Finally, the developing vessels are reinforced by the association with pericytes and smooth muscle cells (Coultas, et al. 2005). In parallel, in a process called lymphangiogenesis, lymphatic endothelial cells, which derive from embryonic veins by sprouting, form primary lymph sacs and the primary lymphatic plexus (Alitalo and Carmeliet 2002).

In the adult, blood vessels are usually quiescent and angiogenesis only occurs in ovary and placenta, as well as during wound healing and repair in response to stimuli as diverse as hypoxia and inflammation. However, in cancer and other disorders, excessive stimulation triggers the "angiogenic switch." Conceivably, angiogenic vessels enhance the supply of oxygen and nutrients to the tumor cells, thereby promoting tumor growth. In addition, lymphatic vessels (which normally maintain the fluid balance) also contribute to the pathogenesis of cancer, by favoring the spreading of metastatic tumor cells from the primary tumor to the regional lymph nodes and, thence, to other nodes and distant organs. Thus, in recent years, research has focused on the molecular mechanisms that regulate angiogenesis and lymphangiogenesis in tumors with the aim of identifying potential molecules for targeted tumor therapy and metastasis prevention (Alitalo, et al. 2005; Carmeliet 2005; Ferrara and Kerbel 2005).

In addition, in the embryo, formation of the organs (including the cardiovascular system) requires adhesive interactions of individual cells with both extracellular matrix and adjacent cells. In turn, these two types of interactions (i.e., cell-matrix and cell-cell adhesion) require several adhesive complexes. Similarly, in the adult,

---

Department of Biochemistry and Molecular Pharmacology, Istituto di Ricerche Farmacologiche Mario Negri, Via La Masa 19, I-20156 Milano (Italy), Phone: +39-02-39014480, Fax: +39-02-3546277, e-mail: bazzoni@marionegri.it

F. Colotta and A. Mantovani (eds.), *Targeted Therapies in Cancer*.
© Springer 2008

cell adhesion molecules play an important role in the angiogenic response and interact functionally with receptors for vascular growth factors. Thus, like signal transduction, endothelial adhesion has also become a productive area of research in the angiogenesis field.

6.2 Pathways in Angiogenesis and Lymphangiogenesis

In this section, we will first describe the vascular micro-environment of tumors. Then, we will outline the major signalling pathways that regulate angiogenesis, with focus on the Vascular Endothelial Growth Factor (VEGF) pathway. Finally, we will briefly analyze the role of lymphangiogenesis in metastasis.

6.2.1 The Vascular Micro-environment of Tumors

Although the vascular endothelial cell is the central player in the process that leads to the neovascularization of tumors, it is now clear that different cell types contribute to this process in a highly coordinated manner (Fig. 6-1). The brief remarks that follow on the so-called angiogenic micro-environment of tumors highlight not only the complexity of the process (Ferrara and Kerbel 2005), but

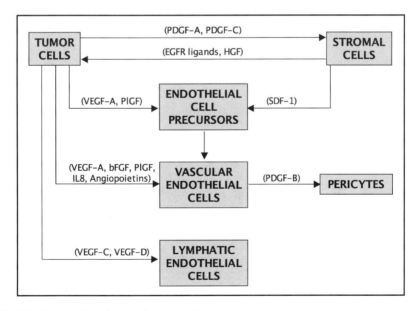

Fig. 6-1 The vascular micro-environment of tumors

also the need of targeting different molecules and cell types for efficient therapeutic results (Yancopoulos, et al. 2000). As a general rule, soluble molecules released by tumor, endothelial and stromal cells (i.e., fibroblasts and immune cells) act as paracrine and autocrine messengers in the communication between the different cell types. First and foremost, tumor cells produce angiogenic factors that stimulate the proliferation and migration of endothelial cells. The most important among such mediators are VEGF-A, basic Fibroblast Growth Factor, Placenta-derived Growth Factor (PlGF), interleukin-8 and the Angiopoietins. Similarly, stromal cells release additional angiogenic factors, such as Stroma-Derived Factor-1 that is mostly produced by fibroblasts and is responsible for the recruitment of bone marrow-derived endothelial cell precursors. Notably, these angiogenic precursors are recruited to the tumor in response to VEGF-A and PlGF as well. In addition, stromal cells produce growth factors for the tumor cells (i.e., Epidermal and Hepatocyte Growth Factors). Conversely, tumor cells release factors, such as Platelet-Derived Growth Factor (PDGF)-A and PDGF-C, that recruit stromal cells. Finally, endothelial cells produce PDGF-B, which reinforces the newly formed vessels by recruiting pericytes.

6.2.2 *The Vascular Endothelial Growth Factor Pathway*

Among the numerous regulators of vasculogenesis and angiogenesis, the VEGF pathway represents one of the most extensively studied systems and one of the most promising targets for anti-angiogenic therapy. Here the molecular features of the VEGF system that are relevant to angiogenesis will be briefly outlined, while a detailed analysis can be found elsewhere (Shibuya and Claesson-Welsh 2006). The VEGF system comprises ligands and receptors that induce partially overlapping functions (Fig. 6-2). At least three different receptors, namely VEGFR1 (Flt-1), VEGFR2 (KDR/Flk-1) and VEGFR3 (Flt-4), are bound by four different ligands. Specifically, VEGF-A binds both VEGFR1 and VEGFR2. At variance, VEGF-B and PlGF bind only VEGFR1. Finally, VEGF-C and VEGF-D bind VEGFR3 (and, albeit much more weakly, VEGFR2). Interestingly, there is also a truncated and soluble form of VEGFR1 that is involved in preeclampsia.

VEGFR1 is a negative regulator of vessel formation in the embryo. Actually, deletion of the *VEGFR1* gene in mice causes embryonic lethality, which is associated with endothelial overgrowth and vascular disorganization (Fong, et al. 1995). The observation that the tyrosine kinase (but not the extracellular) domain of VEGFR1 is dispensable for mediating these abnormal effects suggests that VEGFR1 inhibits vascular development (via its extracellular domain) by trapping VEGF-A. As a consequence, VEGFR1 reduces the interaction of VEGF-A with VEGFR2, thus preventing excessive angiogenic stimulation via VEGFR2 (Hiratsuka, et al. 1998). However, at variance with the vascular endothelium, in monocytes and macrophages the kinase activity of VEGFR1 is required for infiltrating tumors

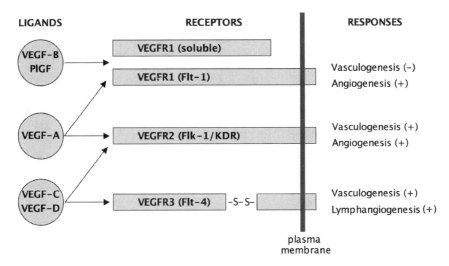

Fig. 6-2 Vascular Endothelial Growth Factor: receptors and ligands

(Lyden, et al. 2001). Given the pro-angiogenic role of inflammatory cells in the tumor micro-environment, it appears that VEGFR1 may be regarded as a positive regulator in the adult (in contrast to its negative role in the embryo).

VEGFR2 is a positive regulator of angiogenesis because its engagement is essential for favoring the differentiation of the endothelial cell progenitors and for sustaining growth and survival of the mature endothelial cells. Actually, deletion of the *VEGFR2* gene in mice causes embryonic lethality, which is associated with lack of vasculogenesis and defective hematopoietic development (Shalaby, et al. 1995). In addition, binding of VEGFR2 stimulates endothelial cell migration in both embryo and adult, which is consistent with its expression at the leading edge of migrating cells (Gerhardt, et al. 2003). Signalling by VEGFR2 allows two types of response that depend upon the tyrosine residue (in the cytoplasmic tail of the receptor) that becomes phosphorylated upon VEGF-A binding. On one side, phosphorylated Y1175 (in the carboxyl terminus) mediates binding of phospholipase C gamma (which results in Protein Kinase C activation, inositol trisphosphate generation and calcium mobilization) and activation of the Raf/MAPK cascade (which results in DNA synthesis). On the other side, phosphorylated Y951 (in the kinase insert) mediates binding of T cell-Specific Adapter, which in turn binds Src, thus regulating actin stress fiber formation and cell migration (Takahashi, et al. 1999).

VEGFR3, which binds VEGF-C and -D (but not VEGF-A and -B), plays an essential role in vascular development, as it mediates remodeling of the primary vascular plexus (Dumont, et al. 1998). Later on, however, the expression of VEGFR3 becomes restricted to the lymphatic endothelial cells, where VEGFR3 is essential for the sprouting of lymphatic vessels from embryonic veins (Karkkainen, et al. 2004).

6.2.3 Other Mediators of Angiogenesis and Lymphangiogenesis

In addition to the VEGF pathway, other signalling pathways play an established role in vascular angiogenesis and are, therefore, potential targets for anti-angiogenic therapy. Specifically, as mentioned above, PDGF-B is involved (via the PDGFR-beta receptor) in recruiting pericytes. This response is important for vessel maturation, as pericytes contribute to the mechanical stability of the capillary wall. Actually, deletion of the *PDGF-B* gene in mice is associated with lack of microvascular pericytes and appearance of capillary aneurysms that rupture at late gestation. Besides being unable to attract PDGFR-beta-positive pericyte progenitors, PDGF-B-deficient endothelial cells become more dependent upon VEGF-A (Lindahl, et al. 1997).

Also the Angiopoietins, which are soluble ligands for the receptor tyrosine kinases Ties, play a role in vessel stabilization. In particular, Angiopoietin-1 is the major agonist for Tie-2, and deletion of either *Angiopoietin-1* or *Tie-2* gene causes similar vascular defects (Suri, et al, 1996). At variance, Angiopoietin-2 has been regarded to as a Tie-2 antagonist, as transgenic over-expression of Angiopoietin-2 in mice disrupts blood vessel formation (Maisonpierre, et al. 1997). More recent evidence, however, suggests that Angiopoietin-2 may also exert (at least in tumor angiogenesis) positive effects, as systemic treatment of tumor-bearing mice with selective blocking agents reduces tumor growth and endothelial cell proliferation. Thus, inhibition of Angiopoietin-2 might represent a promising anti-angiogenic therapy (Oliner, et al. 2004).

Other regulators of angiogenesis of potential pharmacological interest comprise molecules that are normally involved in the regulation of axon guidance, such as members of the neuropilins/semaphorins and ephrins families, as well as Robo/Slit and netrin/Unc-5, even though their role in angiogenesis is not clear yet (Carmeliet and Tessier-Lavigne 2005; Pasquale 2005). Finally, it is worth mentioning that angiogenesis is also inhibited by naturally occurring regulators, such as thombospondin, which is secreted by epithelial cells, as well as the protein fragments endostatin (O'Reilly, et al. 1997), tumstatin (Sund, et al. 2005), vasostatin (Pike, et al. 1998) and vasohibin (Watanabe, et al. 2004).

Finally, as mentioned above, several studies in the past few years have focused on the molecular mechanisms of lymphatic formation, given the importance of lymphangiogenesis in tumor metastasis (Alitalo, et al. 2005; Carmeliet 2005). In addition to the responses initiated by VEGFR3 (upon binding of VEGF-C and -D), other mediators are responsible for the different steps of lymphatic development. Mediators of commitment, which induce budding from the cardinal vein, migration and formation of lymph sacs, comprise the transcription factor Prox1 and the hyaluronan receptor Lymphatic Vessel-1. In addition, formation of the primary lymphatic plexus, as well as separation of blood and lymphatic vasculature requires Syk and SLP76. Finally, other mediators (i.e., Angiopoietin-2, Ephrin B2, Foxc2 and Podoplanin) are required for the maturation of the lymphatic vasculature. Experimental in vivo studies support the notion that blocking lymphangiogenesis might represent an efficient antimetastatic therapy (Achen, et al. 2005). Actually, either an anti-VEGF-D antibody (Stacker, et al. 2001) or a recombinant soluble molecule, which consists of a

VEGFR3/Ig chimera and acts as a trap for VEGF-C and VEGF-D (Karpanen, et al. 2001), significantly reduce the in vivo incidence of lymphatic metastasis.

6.2.4 Angiogenesis Inhibitors

Among the molecular mediators of angiogenesis that represent potential targets for tumor therapy, the most promising results have been obtained using compounds that target the VEGF pathway. The best characterized member of this class is Bevacizumab, which is a humanized monoclonal antibody that recognizes VEGF-A. Its current indication is in patients with metastatic colorectal cancer in association with chemotherapy agents, such as 5-fluororuracil (Hurwitz, et al. 2004). Although Bevacizumab is well tolerated, hemorrhagic and thrombotic complications have been observed. Conceivably, VEGF inhibition decreases the ability of endothelial cells to respond to vascular damage and to prevent tissue factor exposure, thus causing hemorrhage and thrombosis, respectively. Currently, clinical trials are in progress in order to establish efficacy and safety of Bevacizumab in other tumors such as carcinomas, sarcomas and hematological malignancies. Finally, in addition to Bevacizumab, other anti-angiogenesis compounds comprise the oral kinase inhibitors PTK787/ZK222584 (which inhibits VEGFR1, VEGFR2 and PDGF receptor) and ZD6474 (which inhibits VEGFR2, VEGFR3 and Epidermal Growth Factor Receptor). In general, it is commonly perceived that possible benefits will derive from combining these anti-angiogenic agents with more conventional chemotherapy (Zakarija and Soff 2005).

Several anti-angiogenesis agents (such as the VEGF inhibitors) target endothelial cells directly and induce regression of blood vessels. In addition, they also inhibit the recruitment of endothelial precursor cells (Carmeliet 2005). However, alternative anti-angiogenesis strategies may target other cell types in the tumor micro-environment, thereby inhibiting angiogenesis in an endothelial-independent manner. The PDGF inhibitors, for instance, target mural and stromal cells, with the net effect of destabilizing vessels and reducing the release of pro-angiogenic factors. Still other compounds (such as VEGFR1 inhibitors and chemokine antagonists) target primarily hematopoietic cells, thus reducing the infiltration of pro-angiogenic leukocytes and bone marrow-derived precursors. Finally, other treatments (such as chemotherapy and radiotherapy) target cancer cells, thus decreasing the release of angiogenic factors from tumor cells.

6.3 Cell-matrix Adhesion in Angiogenesis: Endothelial Integrins

Transmembrane alpha-beta heterodimers of the integrin superfamily are important mediators of cell adhesion to the extracellular matrix and to the basement membrane (Hynes 2002). Like other cell types, endothelial cells also express a wide variety of

Table 6-1 Integrin expression in vascular endothelial cells

Integrins	Major ligands	Lethality upon gene deletion
β_1 *Integrins*		β_1-/-: + (peri-implantation)
$\alpha_1\beta_1$	Coll I-IV, Lm-1	α_1-/-: -
$\alpha_2\beta_1$	Coll I, Lm-1	α_2-/-: -
$\alpha_3\beta_1$	Lm-5	α_3-/-: + (perinatal)
$\alpha_4\beta_1$	VCAM-1, Fn	α_4-/-: + (embryonic)
$\alpha_5\beta_1$	Fn	α_5-/-: + (embryonic)*
$\alpha_6\beta_1$	Lm-1	α_6-/-: + (perinatal)
$\alpha_v\beta_1$	Vn, Fn	α_v-/-: + (embryonic/perinatal)
β_3 *Integrins*		β_3-/-: -
$\alpha_v\beta_3$	Vn, Fn, MFG-E8	α_v-/-: see above
β_4 *Integrins*		β_4-/-: + (perinatal)
$\alpha_6\beta_4$	Lm-5	α_6-/-: see above
β_5 *Integrins*		β_5-/-: -
$\alpha_v\beta_5$	Vn, MFG-E8	α_v-/-: see above
β_8 *Integrins*		β_8-/-: + (embryonic/perinatal)
$\alpha_v\beta_8$		α_v-/-: see above

Abbreviations: Coll, Collagen; Lm, Laminin; VCAM, Vascular Cell Adhesion Molecule; Fn, Fibronectin; Vn, Vitronectin; MFG-E8, Milk Fat Globule-EGF factor 8.
* Lethality associated with vascular defects (see text for details)

integrins at the cell surface (Table 6-1). In addition, several lines of evidence indicate that the interaction of endothelial integrins with their extracellular ligands favors the formation of blood vessels during vasculogenesis, as well as during developmental and postnatal angiogenesis (Bazzoni, et al. 1999). First, endothelial integrins mediate pro-angiogenic functions, such as migration, proliferation, survival and matrix degradation. Second, integrins interact functionally with the most important angiogenic pathways. In particular, soluble factors (i.e., VEGF, basic Fibroblast Growth Factor and Transforming Growth Factor-beta) modulate integrin expression and function, while integrins, in turn, regulate some growth factor receptors. In this respect it is noteworthy that regulation of vasculogenesis requires modulation of integrin activation by chemoattractant and chemorepulsive cues, such as angiogenic factors and semaphorins (Serini, et al. 2006). Finally, and perhaps most importantly, blocking endothelial integrins with antibodies and peptides strongly affects angiogenesis both in vivo (i.e., rabbit cornea vascularization) and in vitro (i.e., tube formation in collagen gels). In particular, several studies clearly implicate the endothelial integrins alpha-v/beta-3 and alpha-v/beta-5 as angiogenesis mediators and LM609 (a blocking anti-alpha-v/beta-3 antibody) as a potent angiogenesis inhibitor in different experimental models. Actually, a humanized version of LM609 has entered clinical trials (Tucker 2003).

Based on these findings it was hypothesized that null mutations of genes encoding endothelial integrins might arrest vascular development. However, contrary to the expectations, vascular development in the embryo does not appear to depend upon endothelial integrins (with the sole exception of the alpha-5/beta1 integrin and its ligand fibronectin), including the alpha-v/beta-3 and alpha-v/beta5 integrins, which are instead central regulators of postnatal tumor angiogenesis. Actually, deletion of the *alpha-5* integrin gene causes defects (Yang, et al. 1993) that are similar to (or almost as serious as) those induced by the absence of the *fibronectin* gene (Francis, et al. 2002), thus suggesting that the interaction of this extracellular ligand with its integrin receptor is crucial for vessel development during the vasculogenesis stage. In particular, deletion of the *alpha-5* gene causes embryonic lethality due to mesoderm defects with incomplete fusion of yolk sac blood islands into blood vessels.

In contrast, deletion of the *alpha-v* integrin gene causes either embryonic lethality due to placentation defects (in about one-fifth of embryos) or perinatal lethality due to cerebral hemorrhage (Bader, et al. 1998), thus indicating that the vascular abnormalities only occur after the initial phases of vasculogenesis and early angiogenesis (McCarty, et al. 2002). In addition, although the alpha-v chain pairs with different beta chains, the heterodimer responsible for the defects observed in *alpha-v*-deficient mice is likely to be alpha-v/beta-8, as deletion of the *beta-8* gene also causes cerebral hemorrhage (Zhu, et al. 2002), whereas deletion of genes encoding other alpha-v partners, such as *beta-1* (Fassler and Meyer 1995; Stephens, et al. 1995), *beta-3* (Hodivala-Dilke, et al. 1999) and *beta-5* (Reynolds, et al. 2002), has no effects on vascular development. The angiogenic effect of the alpha-v/beta 8-integrin is likely due to the ability of the integrin (expressed on the astrocytes membrane) to bind and activate Transforming Growth Factor-beta in proximity of the basement membrane of brain vessels (Cambier, et al. 2005). Finally, it is noteworthy that mice carrying a null mutation for the *beta-3* and *beta-5* genes (either alone or in combination) not only show normal development of the vessels, but also display enhanced tumor angiogenesis, thus suggesting that alpha-v/beta-3 and alpha-v/beta-5 may even act as negative regulators of vessels formed in the adult (Reynolds, et al. 2002). The discrepancy with the in vitro studies, which show a positive role for these endothelial integrins, is possibly explained by recent findings demonstrating that the anti- or pro-angiogenic outcome depends on whether they are occupied by a soluble (i.e., type IV collagen-derived tumstatin) or an insoluble (i.e., vitronectin) ligand, respectively. Actually, either in the absence of ligand or in the presence of a soluble ligand (i.e., tumstatin), alpha-v/beta-3 recruits Caspase-8 to the plasma membrane and initiates endothelial apoptosis (Stupack, et al. 2001; Maeshima, et al. 2002). In contrast, alpha-v/beta-3 and alpha-v/beta-5 induce pro-angiogenic responses (by amplifying VEGFR2 signaling in response to VEGF), when endothelial cells adhere to insoluble ligands, such as vitronectin (Soldi, et al. 1999) and Milk Fat Globule-8, which is a glycoprotein secreted by vascular endothelial cells (Silvestre, et al. 2005).

Finally, another integrin that has been involved in angiogenesis is the alpha-6/beta-4 integrin, which is a receptor for laminin-5 in the basement membrane. This integrin associates (on the intracellular side) with intermediate filaments, thereby contributing to the formation of adhesive structures known as hemi-desmosomes, even though it retains the ability (upon stimulation) to interact with the actin filaments, thereby contributing to wound healing and local invasion (in normal and transformed cells, respectively). Interestingly, in endothelial cells, alpha-6/beta-4 is expressed only in mature (but not developing) blood vessels, which is consistent with the lack of vascular phenotype in *beta-4*-deficient mouse embryos (van der Neut, et al. 1996). However, mice carrying a targeted mutation in the signaling domain of the beta-4 cytoplasmic tail show defective angiogenesis in response to basic Fibroblast Growth Factor and VEGF (Nikolopoulos, et al. 2004). Thus, these data indicate that alpha-6/beta-4 controls the invasive phase of postnatal angiogenesis in the adult and suggest that inhibiting alpha-6/beta-4 signalling may represent a possible therapeutic strategy.

6.4 Cell-cell Adhesion in Angiogenesis: Endothelial Cadherins

Besides adhering to the extracellular matrix, endothelial cells adhere to each other via intercellular junctions. These types of junctions are composed of transmembrane proteins that mediate homotypic and homophilic (i.e., between similar cells and similar molecules, respectively) adhesion. Inside the cell, the transmembrane proteins bind cytoplasmic molecules, which in turn bind actin filaments. Like epithelial cells, the endothelial cells also form adherens junctions and tight junctions. However, unlike epithelial cells, endothelial cells do not form desmosomes (Bazzoni and Dejana 2004).

At the adherens junctions, the transmembrane proteins belong to the cadherin family, which mediate intercellular recognition and adhesion, in a homophilic and calcium-dependent manner. Endothelial cells specifically express VE (Vascular Endothelial)-cadherin, which is largely responsible for the organization of the adherens junctions. On the cytoplasmic side, cadherins associate with beta- and gamma-catenins, as well as with p120. In addition, beta- and gamma-catenins bind alpha-catenin that, in turn, binds actin, thus establishing a link between adhesion and the cytoskeleton (Dejana, 2004). At variance, at the tight junctions, the transmembrane proteins comprise tetraspan proteins, which are known as occludin and claudins (claudin-5 being an endothelial-specific cadherin), as well single-pass and immunoglobulin-like glycoproteins known as Junctional Adhesion Molecules (Bazzoni 2003). On the cytoplasmic side, tight junctions express a series of different molecules, like Zonula Occludens-1 that, similarly to alpha-catenin, binds actin filaments. Finally, endothelial cells also express other types of adhesive proteins (outside the junctions) such as Platelet Endothelial Cell Adhesion Molecule-1 and Muc-18. While the role of VE-cadherin and adherens junctions in vessel formation is supported by a series of evidence (as detailed below), deletion of genes encoding endothelial components of the tight junctions, such as *occludin* (Saitou, et al. 2000), *Junctional Adhesion Molecule-A* (Cera, et al. 2004) and *claudin-5* (Nitta, et al. 2003), does not result in obvious vascular phenotypes. Nonetheless, deletion of the *claudin-5* gene, while leaving vessel formation unaffected, results in selective defect of the blood-brain barrier for the passage of low molecular weight molecules.

The role of endothelial cadherins in vascular development has been reviewed recently (Cavallaro, et al. 2006; Wallez, et al. 2006). Here we will summarize the evidence pointing out to a role for VE-cadherin in normal and pathological angiogenesis. Concerning embryonic vasculogenesis and angiogenesis, deletion of the *VE-cadherin* gene in mice results in early lethal phenotype. Although VE-cadherin-defective endothelial cells retain the ability to form the primitive vascular plexus, they tend to detach from each other. As a consequence, vessels regress and collapse, thus causing embryonic lethality (Carmeliet, et al. 1999). Concerning postnatal angiogenesis, a VE-cadherin blockade with monoclonal antibodies has been shown to inhibit tumor growth and vascularization in animal models of angiogenesis (Corada, et al. 2001). Clearly, an ideal therapeutic tool should combine maximal efficacy against dynamic junctions of proliferating endothelial cells and minimal effects against stable junctions of quiescent endothelial cells, in order to prevent angiogenesis without enhancing vascular permeability. A series of novel anti-VE-cadherin antibodies has the potential of fulfilling these criteria, thereby further establishing VE-cadherin as a potential target for angiogenesis inhibition in tumors (Cavallaro, et al. 2006).

Given the importance of the VE-cadherin-dependent regulation of vessel formation, research efforts have aimed at defining the effectors of the cadherin. Conceivably, one of the major actions of cadherin (at least in quiescent cells) consists in maintaining adhesive interactions between adjacent cells. However, VE-cadherin (like other adhesive molecules) is also able to impinge on signal transduction, by at least two mechanisms. First, VE-cadherin sequesters beta-catenin at the plasma membrane, thereby preventing its translocation to the nucleus (and its transcription factor activity). Second, VE-cadherin expression attenuates VEGF-induced VEGFR2 phosphorylation in tyrosine, MAP kinase activation and cell proliferation, in a way that requires beta-catenin, and that may contribute to explaining the contact-dependent inhibition of cell proliferation and cell motility. The likely explanation is the following. Upon stimulation with VEGF in contacting cells, VEGFR-2 associates with the VE-cadherin-beta-catenin complex at cell-cell contacts, where it may be inactivated by junctional phosphatases such as DEP-1/CD148. At variance, this phenomenon does not occur in sparse cells (that is, in the absence of cell-cell junctions), thus allowing full activation of the VEGF receptor, as well as cell proliferation and migration (Lampugnani, et al 2003). Finally, VE-cadherin affects the molecular architecture of the endothelial junctions via beta-catenin. Significantly, endothelial cell-specific deletion of the *beta-catenin* gene in mice deeply affects the adherens junctions, with consequent defects in vessel morphology (Cattelino, et al. 2003) and heart defects (Liebner, et al. 2004) that lead to increased permeability and embryonic lethality.

6.5 Conclusions

Anti-angiogenic and anti-lymphangiogenic interventions are promising tools for future tumor therapy. The early results with compounds that interfere with the VEGF signaling pathways are encouraging, even though it is becoming clear that the complex picture of the angiogenic process will require multiple types of therapy (possibly in combination), in order to target different cellular and molecular mediators. Along the same line, targeting endothelial cell adhesive systems (both cell-matrix and cell-cell adhesion) may prove a parallel and perhaps complementary approach for anti-angiogenesis therapy.

Acknowledgements The generous support by AICR (Association International for Cancer Research, St. Andrews, United Kingdom; Grant 04-095) is gratefully acknowledged.

References

Achen, M. G.; B. K. McColl and S. A. Stacker (2005). Focus on lymphangiogenesis in tumor metastasis. Cancer Cell 7: 121–7.
Alitalo, K. and P. Carmeliet (2002). Molecular mechanisms of lymphangiogenesis in health and disease. Cancer Cell 1: 219–27.
Alitalo, K.; T. Tammela and T. V. Petrova (2005). Lymphangiogenesis in development and human disease. Nature 438: 946–53.
Bader, B. L.; H. Rayburn; D. Crowley and R. O. Hynes (1998). Extensive vasculogenesis, angiogenesis, and organogenesis precede lethality in mice lacking all alpha v integrins. Cell 95: 507–19.
Bazzoni, G. (2003). The JAM family of junctional adhesion molecules. Curr Opin Cell Biol 15: 525–30.
Bazzoni, G. and E. Dejana (2004). Endothelial cell-to-cell junctions: molecular organization and role in vascular homeostasis. Physiol Rev 84: 869–901.
Bazzoni, G., E. Dejana and M. G. Lampugnani (1999). Endothelial adhesion molecules in the development of the vascular tree: the garden of forking paths. Curr Opin Cell Biol 11: 573–81.
Cambier, S.; S. Gline; D. Mu; R. Collins; J. Araya; G. Dolganov; S. Einheber; N. Boudreau and S. L. Nishimura (2005). Integrin alpha(v)beta8-mediated activation of transforming growth factor-beta by perivascular astrocytes: an angiogenic control switch. Am J Pathol 166: 1883–94.
Carmeliet, P. (2005). Angiogenesis in life, disease and medicine. Nature 438: 932–6.
Carmeliet, P.; M. G. Lampugnani; L. Moons; F. Breviario; V. Compernolle; F. Bono; G. Balconi; R. Spagnuolo; B. Oostuyse; M. Dewerchin; A. Zanetti; A. Angellilo; V. Mattot; D. Nuyens; E. Lutgens; F. Clotman; M. C. de Ruiter; A. Gittenberger-de Groot; R. Poelmann; F. Lupu; J. M. Herbert; D. Collen and E. Dejana (1999). Targeted deficiency or cytosolic truncation of the VE-cadherin gene in mice impairs VEGF-mediated endothelial survival and angiogenesis. Cell 98: 147–57.
Carmeliet, P. and M. Tessier-Lavigne (2005). Common mechanisms of nerve and blood vessel wiring. Nature 436: 193–200.
Cattelino, A.; S. Liebner; R. Gallini; A. Zanetti; G. Balconi; A. Corsi; P. Bianco; H. Wolburg; R. Moore; B. Oreda; R. Kemler and E. Dejana (2003). The conditional inactivation of the beta-catenin gene in endothelial cells causes a defective vascular pattern and increased vascular fragility. J Cell Biol 162: 1111–22.
Cavallaro, U.; S. Liebner and E. Dejana (2006). Endothelial cadherins and tumor angiogenesis. Exp Cell Res 312: 659–67.
Cera, M. R.; A. Del Prete; A. Vecchi; M. Corada; I. Martin-Padura; T. Motoike; P. Tonetti; G. Bazzoni; W. Vermi; F. Gentili; S. Bernasconi; T. N. Sato A. Mantovani and E. Dejana (2004). Increased DC trafficking to lymph nodes and contact hypersensitivity in junctional adhesion molecule-A-deficient mice. J Clin Invest 114: 729–38.

Corada, M.; F. Liao; M. Lindgren; M. G. Lampugnani; F. Breviario; R. Frank; W. A. Muller; D. J. Hicklin P. Bohlen and E. Dejana (2001). Monoclonal antibodies directed to different regions of vascular endothelial cadherin extracellular domain affect adhesion and clustering of the protein and modulate endothelial permeability. Blood 97: 1679–84.

Coultas, L., K. Chawengsaksophak and J. Rossant (2005). Endothelial cells and VEGF in vascular development. Nature 438: 937–45.

Dejana, E. (2004). Endothelial cell-cell junctions: happy together. Nat Rev Mol Cell Biol 5: 261–70.

Dumont, D. J.; L. Jussila; J. Taipale; A. Lymboussaki; T. Mustonen; K. Pajusola M. Breitman and K. Alitalo (1998). Cardiovascular failure in mouse embryos deficient in VEGF receptor-3. Science 282: 946–9.

Fassler, R. and M. Meyer (1995). Consequences of lack of beta 1 integrin gene expression in mice. Genes Dev 9: 1896–908.

Ferrara, N. and R. S. Kerbel (2005). Angiogenesis as a therapeutic target. Nature 438: 967–74.

Fong, G. H.; J. Rossant; M. Gertsenstein and M. L. Breitman (1995). Role of the Flt-1 receptor tyrosine kinase in regulating the assembly of vascular endothelium. Nature 376: 66–70.

Francis, S. E.; K. L. Goh; K. Hodivala-Dilke; B. L. Bader; M. Stark D. Davidson and R. O. Hynes (2002). Central roles of alpha5beta1 integrin and fibronectin in vascular development in mouse embryos and embryoid bodies. Arterioscler Thromb Vasc Biol 22: 927–33.

Gerhardt, H.; M. Golding; M. Fruttiger; C. Ruhrberg; A. Lundkvist; A. Abramsson; M. Jeltsch; C. Mitchell; K. Alitalo, D. Shima and C. Betsholtz (2003). VEGF guides angiogenic sprouting utilizing endothelial tip cell filopodia. J Cell Biol 161: 1163–77.

Hiratsuka, S.; O. Minowa; J. Kuno T. Noda and M. Shibuya (1998). Flt-1 lacking the tyrosine kinase domain is sufficient for normal development and angiogenesis in mice. Proc Natl Acad Sci U S A 95: 9349–54.

Hodivala-Dilke, K. M.; K. P. McHugh; D. A. Tsakiris; H. Rayburn; D. Crowley; M. Ullman-Cullere; F. P. Ross; B. S. Coller; S. Teitelbaum and R. O. Hynes (1999). Beta3-integrin-deficient mice are a model for Glanzmann thrombasthenia showing placental defects and reduced survival. J Clin Invest 103: 229–38.

Hurwitz, H., L. Fehrenbacher; W. Novotny; T. Cartwright; J. Hainsworth; W. Heim; J. Berlin; A. Baron; S. Griffing; E. Holmgren; N. Ferrara; G. Fyfe; B. Rogers R. Ross and F. Kabbinavar (2004). Bevacizumab plus irinotecan, fluorouracil, and leucovorin for metastatic colorectal cancer. N Engl J Med 350: 2335–42.

Hynes, R. O. (2002). Integrins: bidirectional, allosteric signaling machines. Cell 110: 673–87.

Karkkainen, M. J.; P. Haiko; K. Sainio; J. Partanen; J. Taipale; T. V. Petrova; M. Jeltsch; D. G. Jackson; M. Talikka; H. Rauvala; C. Betsholtz and K. Alitalo (2004). Vascular endothelial growth factor C is required for sprouting of the first lymphatic vessels from embryonic veins. Nat Immunol 5: 74–80.

Karpanen, T.; M. Egeblad; M. J. Karkkainen; H. Kubo; S. Yla-Herttuala; M. Jaattela and K. Alitalo (2001). Vascular endothelial growth factor C promotes tumor lymphangiogenesis and intralymphatic tumor growth. Cancer Res 61: 1786–90.

Lampugnani, M. G.; A. Zanetti; M. Corada; T. Takahashi; G. Balconi; F. Breviario; F. Orsenigo; A. Cattelino; R. Kemler T. O. Daniel and E. Dejana (2003). Contact inhibition of VEGF-induced proliferation requires vascular endothelial cadherin, beta-catenin, and the phosphatase DEP-1/CD148. J Cell Biol 161: 793–804.

Liebner, S.; A. Cattelino; R. Gallini; N. Rudini; M. Iurlaro; S. Piccolo and E. Dejana (2004). Beta-catenin is required for endothelial-mesenchymal transformation during heart cushion development in the mouse. J Cell Biol 166: 359–67.

Lindahl, P.; B. R. Johansson; P. Leveen and C. Betsholtz (1997). Pericyte loss and microaneurysm formation in PDGF-B-deficient mice. Science 277: 242–5.

Lyden, D.; K. Hattori; S. Dias; C. Costa; P. Blaikie; L. Butros; A. Chadburn; B. Heissig; W. Marks; L. Witte; Y. Wu; D. Hicklin; Z. Zhu; N. R. Hackett; R. G. Crystal; M. A. Moore; K. A. Hajjar; K. Manova; R. Benezra and S. Rafii (2001). Impaired recruitment of bone-marrow-derived endothelial and hematopoietic precursor cells blocks tumor angiogenesis and growth. Nat Med 7: 1194–201.

Maeshima, Y.; A. Sudhakar; J. C. Lively; K. Ueki; S. Kharbanda; C. R. Kahn; N. Sonenberg; R. O. Hynes and R. Kalluri (2002). Tumstatin, an endothelial cell-specific inhibitor of protein synthesis. Science 295: 140–3.

Maisonpierre, P. C.; C. Suri; P. F. Jones; S. Bartunkova; S. J. Wiegand; C. Radziejewski; D. Compton; J. McClain; T. H. Aldrich; N. Papadopoulos; T. J. Daly; S. Davis; T. N. Sato and G. D. Yancopoulos (1997). Angiopoietin-2, a natural antagonist for Tie2 that disrupts in vivo angiogenesis. Science 277: 55–60.

McCarty, J. H.; R. A. Monahan-Earley; L. F. Brown; M. Keller; H. Gerhardt; K. Rubin; M. Shani; H. F. Dvorak; H. Wolburg; B. L. Bader A. M. Dvorak and R. O. Hynes (2002). Defective associations between blood vessels and brain parenchyma lead to cerebral hemorrhage in mice lacking alphav integrins. Mol Cell Biol 22: 7667–77.

Nikolopoulos, S. N.; P. Blaikie; T. Yoshioka; W. Guo and F. G. Giancotti (2004). Integrin beta4 signaling promotes tumor angiogenesis. Cancer Cell 6: 471–83.

Nitta, T.; M. Hata; S. Gotoh; Y. Seo; H. Sasaki; N. Hashimoto; M. Furuse and S. Tsukita (2003). Size-selective loosening of the blood-brain barrier in claudin-5-deficient mice. J Cell Biol 161: 653–60.

O'Reilly, M. S.; T. Boehm; Y. Shing; N. Fukai; G. Vasios; W. S. Lane; E. Flynn; J. R. Birkhead; B. R. Olsen and J. Folkman (1997). Endostatin: an endogenous inhibitor of angiogenesis and tumor growth. Cell 88: 277–85.

Oliner, J.; H. Min; J. Leal; D. Yu; S. Rao; E. You; X. Tang; H. Kim; S. Meyer; S. J. Han; N. Hawkins; R. Rosenfeld; E. Davy; K. Graham; F. Jacobsen; S. Stevenson; J. Ho; Q. Chen; T. Hartmann; M. Michaels; M. Kelley; L. Li; K. Sitney; F. Martin; J. R. Sun; N. Zhang; J. Lu; J. Estrada; R. Kumar; A. Coxon; S. Kaufman; J. Pretorius; S. Scully; R. Cattley; M. Payton; S. Coats; L. Nguyen; B. Desilva; A. Ndifor; I. Hayward; R. Radinsky, T. Boone and R. Kendall (2004). Suppression of angiogenesis and tumor growth by selective inhibition of angiopoietin-2. Cancer Cell 6: 507–16.

Pasquale, E. B. (2005). Eph receptor signalling casts a wide net on cell behavior. Nat Rev Mol Cell Biol 6: 462–75.

Pike, S. E.; L. Yao; K. D. Jones; B. Cherney; E. Appella; K. Sakaguchi; H. Nakhasi; J. Teruya-Feldstein; P. Wirth; G. Gupta and G. Tosato (1998). Vasostatin, a calreticulin fragment, inhibits angiogenesis and suppresses tumor growth. J Exp Med 188: 2349–56.

Reynolds, L. E.; L. Wyder; J. C. Lively; D. Taverna; S. D. Robinson; X. Huang; D. Sheppard R. O. Hynes and K. M. Hodivala-Dilke (2002). Enhanced pathological angiogenesis in mice lacking beta3 integrin or beta3 and beta5 integrins. Nat Med 8: 27–34.

Saitou, M.; M. Furuse; H. Sasaki; J. D. Schulzke; M. Fromm; H. Takano; T. Noda and S. Tsukita (2000). Complex phenotype of mice lacking occludin, a component of tight junction strands. Mol Biol Cell 11: 4131–42.

Serini, G.; D. Valdembri and F. Bussolino (2006). Integrins and angiogenesis: a sticky business. Exp Cell Res 312: 651–8.

Shalaby, F.; J. Rossant; T. P. Yamaguchi; M. Gertsenstein; X. F. Wu; M. L. Breitman and A. C. Schuh (1995). Failure of blood-island formation and vasculogenesis in Flk-1-deficient mice. Nature 376: 62–6.

Shibuya, M. and L. Claesson-Welsh (2006). Signal transduction by VEGF receptors in regulation of angiogenesis and lymphangiogenesis. Exp Cell Res 312: 549–60.

Silvestre, J. S.; C. Thery; G. Hamard; J. Boddaert; B. Aguilar; A. Delcayre; C. Houbron; R. Tamarat; O. Blanc-Brude; S. Heeneman; M. Clergue; M. Duriez; R. Merval; B. Levy; A. Tedgui; S. Amigorena and Z. Mallat (2005). Lactadherin promotes VEGF-dependent neovascularization. Nat Med 11: 499–506.

Soldi, R.; S. Mitola; M. Strasly; P. Defilippi; G. Tarone and F. Bussolino (1999). Role of alphavbeta3 integrin in the activation of vascular endothelial growth factor receptor-2. Embo J 18: 882–92.

Stacker, S. A.; C. Caesar; M. E. Baldwin; G. E. Thornton; R. A. Williams; R. Prevo; D. G. Jackson; S. Nishikawa; H. Kubo and M. G. Achen (2001). VEGF-D promotes the metastatic spread of tumor cells via the lymphatics. Nat Med 7: 186–91.

Stephens, L. E.; A. E. Sutherland; I. V. Klimanskaya; A. Andrieux; J. Meneses R. A. Pedersen and C. H. Damsky (1995). Deletion of beta 1 integrins in mice results in inner cell mass failure and peri-implantation lethality. Genes Dev 9: 1883–95.

Stupack, D. G.; X. S. Puente; S. Boutsaboualoy C. M. Storgard and D. A. Cheresh (2001). Apoptosis of adherent cells by recruitment of caspase-8 to unligated integrins. J Cell Biol 155: 459–70.

Sund, M.; Y. Hamano; H. Sugimoto; A. Sudhakar; M. Soubasakos; U. Yerramalla; L. E. Benjamin; J. Lawler; M. Kieran A. Shah and R. Kalluri (2005). Function of endogenous inhibitors of angiogenesis as endothelium-specific tumor suppressors. Proc Natl Acad Sci U S A 102: 2934–9.

Suri, C.; P. F. Jones; S. Patan; S. Bartunkova; P. C. Maisonpierre; S. Davis T. N. Sato and G. D. Yancopoulos (1996). Requisite role of angiopoietin-1, a ligand for the TIE2 receptor, during embryonic angiogenesis. Cell 87: 1171–80.

Takahashi, T., H. Ueno and M. Shibuya (1999). VEGF activates protein kinase C-dependent, but Ras-independent Raf-MEK-MAP kinase pathway for DNA synthesis in primary endothelial cells. Oncogene 18: 2221–30.

Tucker, G. C. (2003). Alpha v integrin inhibitors and cancer therapy. Curr Opin Investig Drugs 4: 722–31.

van der Neut, R.; P. Krimpenfort; J. Calafat, C. M. Niessen and A. Sonnenberg (1996). Epithelial detachment due to absence of hemidesmosomes in integrin beta 4 null mice. Nat Genet 13: 366–9.

Wallez, Y., I. Vilgrain and P. Huber (2006). Angiogenesis: the VE-cadherin switch. Trends Cardiovasc Med 16: 55–9.

Watanabe, K.; Y. Hasegawa; H. Yamashita; K. Shimizu; Y. Ding; M. Abe; H. Ohta; K. Imagawa; K. Hojo; H. Maki, H. Sonoda and Y. Sato (2004). Vasohibin as an endothelium-derived negative feedback regulator of angiogenesis. J Clin Invest 114: 898–907.

Yancopoulos, G. D.; S. Davis; N. W. Gale; J. S. Rudge; S. J. Wiegand and J. Holash (2000). Vascular-specific growth factors and blood vessel formation. Nature 407: 242–8.

Yang, J. T.; H. Rayburn and R. O. Hynes (1993). Embryonic mesodermal defects in alpha 5 integrin-deficient mice. Development 119: 1093–105.

Zakarija, A. and G. Soff (2005). Update on angiogenesis inhibitors. Curr Opin Oncol 17: 578–83.

Zhu, J.; K. Motejlek; D. Wang; K. Zang, A. Schmidt and L. F. Reichardt (2002). beta8 integrins are required for vascular morphogenesis in mouse embryos. Development 129: 2891–903.

7
Developing T-Cell Therapies for Cancer in an Academic Setting

Malcolm K. Brenner

7.1 Introduction

The development of targeted drug therapies for cancer has been a long-term ambition of clinicians and researchers alike. While most effort has been expended on identifying small molecule therapeutics, the immune system can be manipulated to provide biological components that will be the most targeted of all. Exploitation of the humoral immune system by manufacture of monoclonal antibodies has already been proven highly effective for treatment of many tumors.[1] The successful manipulation of the cellular component of the immune response, however, has been somewhat slower. In part, this is because of the complexity of developing and manufacturing the cellular components, and in part also, because our understanding of cellular physiology and function has been less extensive and detailed. Immunotherapies based on dendritic cells appear promising and adoptive therapy with T lymphocytes is also finally gaining traction. The use of T cells in cancer has a number of potential advantages:

- They have high targeting specificity
- They are capable of recognizing internal antigens if these are processed and presented on the tumor cell surface
- They have a good biodistribution and are able to actively traffic through multiple tissue planes
- They kill their target cells through a wide range of effector mechanisms so that resistance to all of these in a single cancer cell is unlikely
- They are self-amplifying – so that a small number of cells administered initially has the potential to expand to numbers sufficient to eradicate even bulky tumors

Over the past few years these potential benefits of T lymphocytes as anticancer agents have been progressively validated – initially by using T cells to treat relapsed leukemia following stem cell transplantation (donor lymphocyte infusions), and subsequently with the administration of Epstein-Barr virus specific and melanoma

---

Center for Cell and Gene Therapy. Baylor College of Medicine. The Methodist Hospital, Texas Children's Hospital. mbrenner@bcm.edu

specific cytotoxic T lymphocytes to patients with lymphoma, nasopharyngeal carcinoma and melanoma.[2-16]

In this article, I will briefly illustrate the clinical approaches now being used, using as an example cytotoxic T lymphocytes directed to Epstein-Barr virus expressing tumors. I will also explain how this therapeutic approach can be explored in an academic setting before developing commercially viable treatment models.

7.2 Epstein - Barr Virus

Epstein-Barr virus affects 90 percent of the population and acute infection is followed by a life long persistent/latent state. During latency, there is expression of a limited array of viral latency proteins (Fig. 7-1). Infection is usually benign, but the latent virus may be associated with malignant transformation in B or T lymphocytes or in epithelial cells. There are three latency states, which are associated with progressively greater expression of progressively more immunogenic EBV associated antigens. The most immunogenic latency state is classed as Type III, and is associated with immunoblastic lymphoma. The high level of expression of immunogenic latency antigens, such as EBNA 2 and 3, means that these tumors can only occur in patients who are severely immunocompromised, for example, after organ grafting or because of AIDS.

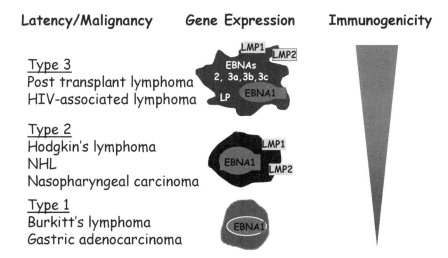

Fig. 7-1 Epstein-Barr Virus Latency. After acute infection, EBV may cause tumors that can be classified by the pattern of viral latency antigens expressed. Type III tumors are the most immunogenic; Type I, the least. Type III latency tumors occur only in immuno-compromised hosts, whereas other latency tumors may occur even in the immuno-competent

7.3 Treatment of Type 3 Latency Tumors (Immunoblastic Lymphoma)

We reasoned that if we could generate cytotoxic T lymphocytes (CTLs) ex vivo that were specific for these latency antigens, we would be able to reconstitute the recipients' immunity to the virus and cure their lymphomas. Accordingly, we made cytotoxic T lymphocytes ex vivo and genetically marked them with a retroviral vector encoding the neomycin phosphotransferase gene.[17] This marker gene allowed us to track the cells in vivo, and determine their survival, expansion and tissue trafficking (Fig. 7-2a-c).[5,6] We found that when the cells were given to patients after stem cell transplantation, they expanded 1,000- to 10,000-fold within one week of administration. They were able to traffic to areas of the tumor, and in five out of six patients, could eradicate bulky malignant disease, even when given in numbers as small as $10^7 m^2$.[9] Of more than 70 patients who received EBV CTLs as prophylaxis, none have developed lymphoproliferative disease, as opposed to 12 percent of our control transplant group.[6,9] Similar effectiveness has been observed in patients with EBV lymphoma after solid organ transplantation.[12]

7.4 Use of EBV CTLs to Treat Malignancy in the Immuno-competent Host

While these results were encouraging, they have little to do with the exigencies of treating cancer in an immuno-competent individual. In the presence of a functioning immune system, potentially immunogenic tumors have to express weak antigens and to develop immune evasion strategies to subvert any responses made by the host. EBV Type II latency tumors are good examples of this process, expressing antigens that are poorly stimulatory to host T cells (such s LMP1 and LMP2), and producing a range of soluble factors (such as TGF-β) that directly inhibit T cell growth and activity, or that attract T regulatory cells to the site of the tumor, both of which serve to limit the potency of the immune response.

As a first step we treated Type II latency Hodgkin's Disease – or nasopharyngeal carcinoma – with EBV-CTLs generated in the same way as the treatment of Type III latency.[7] Although only a very small proportion (less than 1 percent) of the

Fig. 7-2 EBV-Cytotoxic T lymphocytes (CTL) expand and persist after infusion and traffic to tumor sites. Patients at high risk of EBV-lymphoproliferative disease, or who were suffering this complication, were treated with infusion of donor CTLs after allogeneic hemopoietic stem cell transplantation (HSCT). a) There is a 1,000- to 10,000-fold expansion of marker gene positive CTLs within seven days of injection (pre- versus post-PCR for *neo* transgene on five patients). b) Marked cells can be detected in the peripheral blood of patients for 9+ years - plot shows when marked cells were detected in each patient. Time given as days after HSCT. c) In situ PCR of EBV+ lymphoma showing marker gene positive CTL

7 Developing T-Cell Therapies for Cancer in an Academic Setting

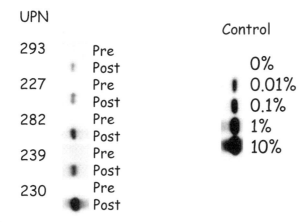

a.

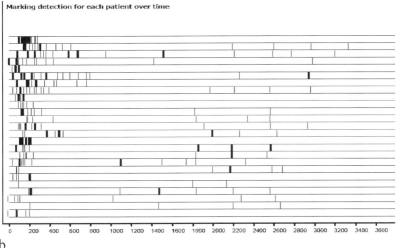

b.

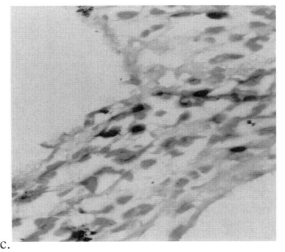

c.

resultant CTLs would be expected to be specific for the weak EBV antigens (LMP1 and LMP2) actually expressed by the malignancy, we reasoned that in the presence of tumor-associated LMP1 and LMP2 in vivo, we would obtain expansion of the subpopulation of CTLs. Indeed, our results showed promise in both Hodgkin's Disease and nasopharyngeal carcinoma.[7] In Hodgkin's Disease there was in vivo expansion of LMP2 and LMP1 specific T cells, while tumor biopsies showed that there was trafficking to the site of malignant disease.[7] Less in vivo expansion occurred in nasopharyngeal carcinoma patients, but trafficking to sites of the tumor still occurred.[8] Of 16 patients with Hodgkin's Disease and nasopharyngeal carcinoma, 5 (31 percent) went into complete remission, and further patients developed a partial remission, so that the PR and CR rate was 68 percent. Since all these patients had disease that was relapsed or resistant after extensive conventional chemo/radiotherapy, these are encouraging results.

7.5 Increasing the Frequency of Tumor Reactive T-Cells

In an effort to improve CTL potency, our next protocol used genetic manipulation of antigen-presenting cells to increase the proportion of infused cells that were specific for the tumor-associated antigen, LMP2. We modified our standard manufacturing approach, and transduced the antigen-presenting cells we had been using with an adenoviral vector encoding LMP2[18] (Fig. 7-3). This resulted in overexpression of the weak tumor antigen and increased the proportion of cytotoxic T lymphocytes that recognized LMP2-associated epitopes 1,000-fold or more. There was a corresponding decline in the frequency of CTLs recognizing latency antigens (such as EBNA 2b), which are normally immuno-dominant, but which are absent on tumor cells and are, thus, irrelevant for CTLs targeting Hodgkin's disease and nasopharyngeal carcinoma (Fig. 7-4). We have given LMP2-specific CTLs to six patients with LMP2-positive lymphoma, and five have had a tumor response, which was complete in four. It is our hope that by incorporating the LMP1 antigen into our adenoviral vector we will broaden the range of tumor antigens that are recognized and further augment antitumor activity. This combination study with LMP1 and LMP2 begins accrual in early 2008.

7.6 Targeting Tumor Stem Cells

Since EBV plays an essential role in the generation of EBV-associated malignancies, antigens associated with the virus are likely to be present even in the early stem cell population. For many other malignancies, however, target antigens that are present in the bulk "late" tumor population may be absent from any progenitor population the tumor contains. Not all tumor-associated antigens are directly involved with the malignant process and so may not be expressed throughout

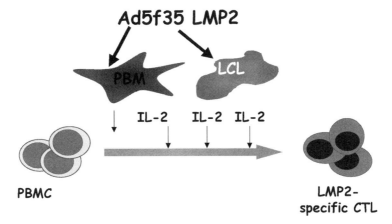

Bollard et al, J Immunother 2004

Fig. 7-3 Adenoviral vector containing the LMP-2 transgene was used to infect dendritic cells and EBV-LCL, and thereby cause them to over-express the normally immuno-subdominant antigen LMP2 at high levels, thereby generating a high proportion of LMP-2 specific CTL (see Fig. 7-4) levels

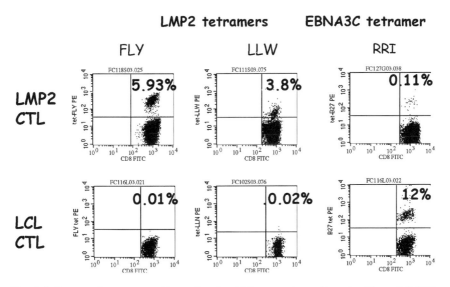

Fig. 7-4 Pattern of tetramer-positive cells after CTL generation with and without AdLMP2. Upper row shows the percentage of CTL that bind two LMP tetramers (FLY and LLW) and one EBNA3C tetramer (RRI) when AdLMP2 was used to over-express the weak tumor-associated antigen LMP2 in the antigen presenting cells. Lower row shows the percentage of CTL that bind these tetramers when we substitute lymphoblastoid cell lines expressing limited amounts of the LMP2 antigen, but plentiful amounts of immuno-dominant antigens such as EBNA3C. The upper row shows a 100- to 500-fold increase in the percentage of cells expressing the tumor-associated antigen LMP2 compared to the lower row, and a 100-fold decrease in cells recognizing EBNA3C, which is lacking on Class II latency tumors and is, therefore, irrelevant for immuno-therapy of these malignancies

tumor maturation. Immuno-therapies targeting such antigens will only be able to modulate the mature component of the tumor and will not eradicate the underlying disease. Fortunately, many immunotherapies can be directed to the precursor population as well.

7.7 CD40L/IL2 B-CLL Tumor Vaccines

We are treating patients with chronic B-lymphocytic leukemia using a vaccine in which human CD40 ligand and human IL2 are expressed transgenically.[19] CD40 ligand has a number of immuno-stimulatory effects. It activates and matures dendritic cells, co-stimulates T cells, and it stimulates the B leukemia cells themselves through their CD40 receptor. B-CLL activation is, in turn, accompanied by up-regulation of MHC Class II molecules and of co-stimulator molecules CD80 and CD86 on the malignant cells, improving their ability to present their own tumor-associated antigens. In animal studies we found that the effects of CD40 ligand were potentiated if the vaccine also expressed Interleukin 2, because this cytokine amplified immune responses made to the CD40L tumor vaccine.[20] We used this vaccine to treat eight patients between 46 and 71 years of age, looking for evidence of toxicity, an immune response to the tumor, and a tumor response.

No severe systemic toxicity was seen after a total of 49 injections in the eight patients, although there was significant local inflammation at the injection site and some patients had flu-like symptoms, which persisted for two to three days.[19] Biopsies of the immunization site showed profound perivascular infiltration with CD3, CD4 and CD8+ cells, but no residual tumor. In most patients we also found evidence of systemic anti-B-CLL immunity, which appeared after three injections and was maximal after six. The patients' T cells generated increased numbers of Elispots in response to autologous, but not allogeneic CLL cells, and there was an antibody response against the tumor cells themselves. However, this immune response was short lived and, two to three months after termination of immunization, the responsiveness had returned to pre-treatment levels. No patient within the first six months showed any response in peripheral blood B cell counts.[19]

7.8 Targeting Side Population Tumor Cells with Tumor Vaccines

One explanation for the failure of immunization to produce a clinical anti-B-CLL response, despite the presence of antitumor immunity, is that we have failed to eradicate a progenitor cell population. It is possible to define a subset of normal hematopoietic stem cells called Side Population (SP) cells, because of their appearance on FACS analysis with Hoechst dyes (Fig. 7-5a).[21] These cells express a high

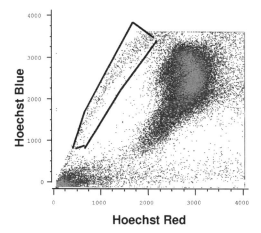

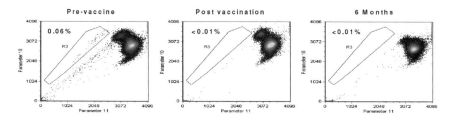

Fig. 7-5 Normal and malignant Side Population (SP) Cells. a) Normal side population cells (gating shown) with high Hoechst dye efflux. b) Malignant SP cells from patient with B-CLL. After vaccination with hCD40L and hIL2 expressing autologous B-CLL, most patients had a temporary decline in SP cells, which then reappeared. In this patient, however, the cells disappeared during vaccination and did not return

level of multiple ABC transporter proteins, and rapidly pump out Hoechst dyes to produce the characteristic appearance seen in Fig. 7-5a. Although the physiological purpose of these pumps is unknown, transplantation of SP cells from the marrow of many species has convincingly shown that these cells are a highly proliferative population, with multi-lineage potential, and that a single SP cell can effectively reconstitute an animal with all hemopoietic components.

More recent study of several malignant tumors has shown that these too could contain a side population of malignant cells. Thus, we were able to show malignant SP cells in bone marrow from patients with acute myeloid leukemia and in patients with the solid tumors neuroblastoma, embryonal and breast and lung cancer.[22] The existence of a tumor SP cell may have particular significance clinically, because the ABC transporters they express pump out not just Hoechst dyes, but also many cytotoxic drugs that are commonly used for treatment. Hence, an SP progenitor population may be highly resistant to agents that can eradicate mature tumor cells, dooming the therapy to failure.

Malignant cells from B-CLL patients also contain a Side Population. Like other tumor SP cells, B-CLLL SP have a more primitive phenotype than the non-SP B-CLL, and over-express ABCG2 and other transporter proteins. When we immunized patients with our CD40L and IL2 vaccine we found an immune response occurred against the SP cells, and that this population was profoundly diminished. However, the cells rapidly rebounded once immunization ceased, and their removal was incomplete. In one patient, however, the side population cells disappeared completely during immunization, and did not recover when immunization was halted (Fig. 7-5b). As described above, this patient - like all others on study - had no change at all in peripheral blood counts during the initial six-month evaluation period. On more prolonged follow-up, however, his peripheral blood B-CLL cells progressively diminished in number, as did his splenomegaly and lymphadonopathy. By 18 months following treatment, his peripheral blood BCLL counts had fallen to less than 1/25th of their starting value. While this is only a single patient, the results are, nonetheless, consistent with the eradication of a tumor stem cell population and subsequent disappearance of the (long-lived) "mature" cells of the tumor, because they can no longer be replenished from a stem cell pool.

To exploit the above benefits of vaccination more effectively, we have restarted our studies with CD40L and IL2 modified vaccines, using a more prolonged period of immunization.

7.9 How Does Immunization with Mature Tumor Cells Induce an Antitumor-Stem-Cell Response?

We immunize patients with the bulk tumor cell population, only 0.5 percent or less of which are SP cells. Why, then, is the most effective immune response observed against such a minority cell population? As mentioned in the preceding section, SP tumor cells express more primitive antigens than the mature non-SP population. In B-CLL these antigens include proteins associated with highly proliferative cells in general, such as hTERT and survivin, and the "stemness" associated proteins $P21^{cip/waf1}$ and BMI-1. We have found that when transgenic CD40 ligand in the tumor vaccine activates B-CLL through their CD40 receptor, there is subsequent up-regulation of expression of many genes in the bulk population that are normally only expressed by B-CLL SP cells. For example, both hTERT and survivin are increased once non-SP cells have encountered CD40L. In other words, CD40L tumor vaccines make mature B-CLL cells more "SP-like," allowing them to act as an immunogen which produces a response primarily targeted to the shared genes expressed by both CD40-activated BCLL cells and the SP BCLL cell population. These results and analyses are highly preliminary, but if confirmed, they will provide a mechanism by which even rare tumor stem cells can be effectively targeted, by rendering the bulk tumor cells more primitive in phenotype and, thus, closer to the optimal target stem cell population.

7.10 Implementation of T-Cell Therapeutics in an Academic Environment

Amongst the greatest challenges to the development of targeted cell therapies are the complexity of the agents and the need for iterative studies that move from laboratory to clinic and back again, testing and manipulating each of the many components of the cell therapeutic to optimize the outcome. As a consequence, most T-cell therapeutics of the type described here cannot readily fit the classical pharmaceutical model. They are instead much closer in concept to the distributive models of surgery, radiotherapy, or hemopoietic stem cell transplantation, in which therapies are assembled and administered at the point of care by a team of skilled workers. Because these new therapies afford the potential to effectively treat "unmet needs" in serious illness, and also promise to have a lower incidence of adverse effects, their cost-effectiveness should be high. Their integration with current industry and medical practices will, nonetheless, require retraining and reallocation of financial resources. The key to achieving this shift in training and finance is the development of approaches that have both a high and a predictable success rate. Unlike a pharmaceutical agent for cancer that can be successfully marketed with a 10 to 12 percent response (not cure) rate, cellular therapy, with its need for large and prolonged institutional investment in resources, personnel, and training, will have a much higher initial energy barrier for implementation. Nonetheless, there are an increasing number of patients who were resistant to conventional therapeutics who are now long-term survivors of cell therapies of the types described above. Fig. 7-6 shows that overall, 40 percent of patients with resistant/relapsed EBV-associated malignancies have complete tumor responses to EBV-CTL therapy, and almost two-thirds have an objective benefit. We are applying for orphan drug status for Epstein-Barr virus-specific cytotoxic lymphocytes to enable us to recover costs of manufacture and administration, and as we progressively improve our approaches for other malignancies in the manner outlined above, these therapies should progressively be brought into wider clinical practice. Hence, we can reasonably look forward to cellular immunotherapy joining new small molecules and monoclonal antibodies as targeted therapeutics for cancer.

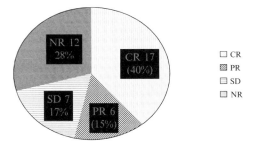

Fig. 7-6 Overall tumor response rate after administration of EBV-specific cytotoxic T lymphocytes (CTL). Forty-two patients with active relapsed or resistant EBV+ lymphoma or nasopharyngeal cancer were treated. Numbers (and percentages) of complete and partial response (CR,PR) are shown, as are the patients with stable disease (SD), or who had no response (NR)

7.11 Summary

T-cell immunotherapies have great potential because of their exquisite targeting ability and high efficacy, but are more complex to implement since they do not readily fit the pharmaceutical model. Nonetheless, recent improvements in gene-transfer technology allow T cells directed to tumor-associated antigens to effectively eradicate even bulky malignancies. T lymphocytes also have the potential to target tumor stem cells, provided the target antigens are appropriately identified and expressed on the components used to generate the effector T-cells. Numerous problems remain before these approaches can be widely implemented, but as the success rate increases, willingness to commit the necessary resources will correspondingly increase.

References

1. Reichert, J. M.; Rosensweig, C. J.; Faden, L. B.; Dewitz, M. C. Monoclonal antibody successes in the clinic. Nat Biotechnol. 2005;23:1073–1078.
2. Porter, D. L., Antin, J. H. The graft-versus-leukemia effects of allogeneic cell therapy. Annu Rev Med. 1999;50:369–386.
3. Kolb, H. J.; Schmid, C.; Barrett, A. J.; Schendel, D. J. Graft-versus-leukemia reactions in allogeneic chimeras. Blood. 2004;103:767–776.
4. O'Reilly, R. J.; Small, T. N.; Papadopoulos, E.; Lucas, K.; Lacerda, J.; Koulova, L. Adoptive immuno-therapy for Epstein-Barr virus-associated lymphoproliferative disorders complicating marrow allografts. Springer Semin Immunopathol. 1998;20:455–491.
5. Heslop, H. E.; Ng, C. Y. C.; Li, C.; Smith, C. A.; Loftin, S. K.; Krance, R. A.; Brenner, M. K.; Rooney, C. M. Long-term restoration of immunity against Epstein-Barr virus infection by adoptive transfer of gene-modified virus-specific T lymphocytes. Nature Medicine. 1996;2:551–555.
6. Rooney, C. M.; Smith, C. A.; Ng, C. Y. C.; Loftin, S. K.; Sixbey, J. W.; Gan, Y-J.; Srivastava, D-K; Bowman, L. C; Krance, R. A.; Brenner, M. K.; Heslop, H. E. Infusion of cytotoxic T cells for the prevention and treatment of Epstein-Barr virus-induced lymphoma in allogeneic transplant recipients. Blood. 1998;92:1549–1555.
7. Bollard, C. M.; Aguilar, L.; Straathof, K. C.; Gahn, B.; Huls, M. H.; Rousseau, A.; Sixbey, J.; Gresik, M. V.; Carrum, G.; Hudson, M.; Dilloo, D.; Gee, A.; Brenner, M. K.; Rooney, C. M.; Heslop, H. E. Cytotoxic T Lymphocyte Therapy for Epstein-Barr Virus+ Hodgkin's Disease. J Exp Med. 2004;200:1623–1633.
8. Straathof, K. C.; Bollard, C. M.; Popat, U.; Huls, M. H.; Lopez, T.; Morriss, M. C.; Gresik, M. V.; Gee, A. P.; Russell, H. V.; Brenner, M. K.; Rooney, C. M.; Heslop, H. E. Treatment of Nasopharyngeal Carcinoma with Epstein-Barr Virus-specific T Lymphocytes. Blood. 2005;105:1898–1904.
9. Gottschalk, S.; Rooney, C. M.; Heslop, H. E. Post-Transplant Lymphoproliferative Disorders. Annu Rev Med. 2005;56:29–44.
10. Comoli, P.; Pedrazzoli, P.; Maccario, R.; Basso, S.; Carminati, O.; Labirio, M.; Schiavo, R.; Secondino, S.; Frasson, C.; Perotti, C.; Moroni, M.; Locatelli, F.; Siena, S. Cell Therapy of Stage IV Nasopharyngeal Carcinoma With Autologous Epstein-Barr Virus-Targeted Cytotoxic T Lymphocytes. J Clin Oncol. 2005.
11. Comoli, P.; Labirio, M.; Basso, S.; Baldanti, F.; Grossi, P.; Furione, M.; Vigano, M.; Fiocchi, R.; Rossi, G.; Ginevri, F.; Gridelli, B.; Moretta, A.; Montagna, D.; Locatelli, F.; Gerna, G;

Maccario, R. Infusion of autologous Epstein-Barr virus (EBV)-specific cytotoxic T cells for prevention of EBV-related lymphoproliferative disorder in solid organ transplant recipients with evidence of active virus replication. Blood. 2002;99:2592–2598.
12. Savoldo, B.; Goss, J. A.; Hammer, M. M.; Zhang, L.; Lopez, T.; Gee, A. P.; Lin, Y. F.; Quiros-Tejeira, R. E.; Reinke, P.; Schubert, S; Gottschalk, S.; Finegold, M. J.; Brenner, M. K.; Rooney, C. M.; Heslop, H. E. Treatment of solid organ transplant recipients with autologous Epstein Barr virus-specific cytotoxic T lymphocytes (CTLs). Blood. 2006;108:2942–2949.
13. Morgan, R. A.; Dudley, M. E.; Wunderlich, J. R.; Hughes, M. S.; Yang, J. C.; Sherry, R. M.; Royal, R. E.; Topalian, S L.; Kammula, U. S.; Restifo, N. P.; Zheng, Z.; Nahvi, A.; de Vries, C. R.; Rogers-Freezer, L. J.; Mavroukakis, S. A.; Rosenberg, S. A. Cancer Regression in Patients After Transfer of Genetically Engineered Lymphocytes. Science. 2006;314:126–129.
14. Dudley, M. E.; Wunderlich, J. R.; Robbins, P. F.; Yang, J. C.; Hwu, P.; Schwartzentruber, D. J.; Topalian, S. L.; Sherry, R.; Restifo, N. P.; Hubicki, A. M.; Robinson, M. R.; Raffeld, M.; Duray, P.; Seipp, C. A.; Rogers-Freezer, L.; Morton, K. E.; Mavroukakis, S. A.; White, D. E.; Rosenberg, S. A. Cancer regression and autoimmunity in patients after clonal repopulation with antitumor lymphocytes. Science. 2002;298:850–854.
15. Dudley, M. E.; Wunderlich, J. R.; Robbins, P. F.; Yang, J. C.; Hwu, P.; Schwartzentruber, D. J.; Topalian, S. L.; Sherry, R.; Restifo, N. P.; Hubicki, A. M.; Robinson, M. R.; Raffeld, M.; Duray, P.; Seipp, C. A.; Rogers-Freezer, L.; Morton, K. E.; Mavroukakis, S. A.; White, D. E.; Rosenberg, S. A. Cancer regression and autoimmunity in patients after clonal repopulation with antitumor lymphocytes. Science. 2002;298:850–854.
16. Yee, C.; Thompson, J. A.; Byrd, D.; Riddell, S. R.; Roche, P.; Celis, E.; Greenberg, P. D. Adoptive T cell therapy using antigen-specific CD8+ T cell clones for the treatment of patients with metastatic melanoma: in vivo persistence, migration, and antitumor effect of transferred T cells. Proc Natl Acad Sci U S A. 2002;99:16168–16173.
17. Smith, C. A.; Ng, C. Y. C.; Heslop, H. E.; Holladay, M. S.; Richardson, S.; Turner, E. V.; Loftin, S. K.; Li, C.; Brenner, M. K.; Rooney, C. M. Production of genetically modified EBV-specific cytotoxic T cells for adoptive transfer to patients at high risk of EBV-associated lymphoproliferative disease. J Hemather. 1995;4:73–79.
18. Bollard, C. M.; Straathof, K. C.; Huls, M. H.; Leen, A.; Lacuesta, K.; Davis, A.; Gottschalk, S.; Brenner, M. K.; Heslop, H. E.; Rooney, C. M. The generation and characterization of LMP2-specific CTLs for use as adoptive transfer from patients with relapsed EBV-positive Hodgkin's disease. J Immunother. 2004;27:317–327.
19. Biagi, E.; Rousseau, R.; Yvon, E.; Schwartz, M.; Dotti, G.; Foster, A.; Havlik-Cooper, D.; Grilley, B. Gee, A.; Baker, K.; Carrum, G.; Rice, L.; Andreeff, M.; Popat, U.; Brenner, M. Responses to human CD40 ligand/human interleukin-2 autologous cell vaccine in patients with B-cell chronic lymphocytic leukemia. Clin Cancer Res. 2005;11:6916–6923.
20. Dilloo, D.; Brown, M.; Roskrow, M.; Zhong, W.; Holden, W.; Holladay, M.; Brenner, M. CD40 ligand induces an anti-leukemia immune response in vivo. Blood. 1997;90:1927–1933.
21. Goodell, M. A.; Rosenzweig, M.; Kim, H.; Marks, D. F.; DeMaria, M.; Paradis, G.; Grupp, S. A.; Sieff, C. A.; Mulligan, R. C.; Johnson, R. P. Dye efflux studies suggest that hematopoietic stem cells expressing low or undetectable levels of CD34 antigen exist in multiple species. Nature Medicine. 1997;3:1337–1345.
22. Hirschmann-Jax, C.; Foster, A. E.; Wulf, G. G.; Nuchtern, J. G.; Jax, T. W.; Gobel, U.; Goodell, M. A.; Brenner, M. K. A distinct "side population" of cells with high drug efflux capacity in human tumor cells. Proc Natl Acad Sci U S A. 2004;101:14228–14233.

8
Anticancer Cell Therapy with TRAIL-Armed CD34[+] Progenitor Cells

Carmelo Carlo-Stella[1], Cristiana Lavazza[2], Antonino Carbone[3], and Alessandro M. Gianni[4]

8.1 Introduction

Dysregulated apoptosis plays a key role in the pathogenesis and progression of neoplastic disorders, allowing tumor cells to survive beyond their normal life-span, and to eventually acquire chemo-radioresistance (Laconi, Pani and Farber, 2000; Pommier, Sordet, Antony, Hayward and Kohn 2004). Thus, apoptotic pathways represent attractive therapeutic targets for restoring apoptosis sensitivity of malignant cells, or activating agonists of apoptosis. To modulate apoptotic genes and proteins, several strategies can be envisaged which target either the mitochondria-dependent or the death receptor-dependent pathways of apoptosis (Waxman and Schwartz 2003). Due to the ability of death receptor ligands to induce death in susceptible cell types, there has been considerable interest in the therapeutic potential of these cytokines as anticancer agents. Death receptor ligands of the tumor necrosis factor α (TNFα) superfamily are type II transmembrane proteins that signal to target cells upon cell-cell contact, or after protease-mediated release to the extracellular space (Ashkenazi 2002). Four members of this family – including Fas ligand (FasL), TNFα, TL1A, and tumor necrosis factor-related apoptosis-inducing ligand (TRAIL) – stand out because of their ability to induce cell death (Wiley, Schooley, Smolak, Din, Huang, Nicholl, Sutherland, Smith, Rauch and Smith 1995; Wajant 2003).

TRAIL, in its soluble form, is emerging as an attractive anticancer agent due to its cancer cell-specificity and potent antitumor activity. TRAIL signals by interacting with its five receptors, including the two agonistic receptors TRAIL-R1 (Pan, O'Rourke, Chinnaiyan, Gentz, Ebner, Ni and Dixit 1997) and TRAIL-R2 (Walczak,

---

[1] University of Milano and Istituto Nazionale Tumori, Department of Medical Oncology, Milano, Italy, carmelo.carlostella@unimi.it

[2] University of Milano and Istituto Nazionale Tumori, Department of Medical Oncology, Milano, Italy, cristiana.lavazza@unimi.it

[3] Istituto Nazionale Tumori, Department of Pathology, Milano, Italy, antonino.carbone@istitutotumori.mi.it

[4] University of Milano and Istituto Nazionale Tumori, Department of Medical Oncology, Milano, Italy, Alessandro.Gianni@unimi.it

Degli-Esposti, Johnson, Smolak, Waugh, Boiani, Timour, Gerhart, Schooley, Smith, Goodwin and Rauch 1997), and the three antagonistic receptors (Sheridan, Marsters, Pitti, Gurney, Skubatch, Baldwin, Ramakrishnan, Gray, Baker, Wood, Goddard, Godowski and Ashkenazi 1997) TRAIL-R3 (Pan, Ni, Wei, Yu, Gentz and Dixit 1997), TRAIL-R4 (Degli-Esposti, Dougall, Smolak, Waugh, Smith and Goodwin 1997), and osteoprotegerin (OPG) (Emery, McDonnell, Burke, Deen, Lyn, Silverman, Dul, Appelbaum, Eichman, DiPrinzio, Dodds, James, Rosenberg, Lee and Young 1998). Both TRAIL-R1 and TRAIL-R2 are type I transmembrane proteins containing a cytoplasmic death domain (DD) motif that engage apoptotic machinery upon ligand binding (Almasan and Ashkenazi 2003), whereas the other three receptors either act as decoys or transduce antiapoptotic signals (Wang and El-Deiry 2003). In vitro several sets of evidence demonstrate that TRAIL selectively induces apoptosis in a variety of transformed cell lines (Pitti, Marsters, Ruppert, Donahue, Moore and Ashkenazi 1996; Mariani, Matiba, Armandola and Krammer 1997; Almasan, et al. 2003); in vivo administration of TRAIL to mice exerts a remarkable activity against various tumor xenografts (Ashkenazi, Pai, Fong, Leung, Lawrence, Marsters, Blackie, Chang, McMurtrey, Hebert, DeForge, Koumenis, Lewis, Harris, Bussiere, Koeppen, Shahrokh and Schwall 1999; Walczak, Miller, Ariail, Gliniak, Griffith, Kubin, Chin, Jones, Woodward, Le, Smith, Smolak, Goodwin, Rauch, Schuh and Lynch 1999; Mitsiades, Treon, Mitsiades, Shima, Richardson, Schlossman, Hideshima and Anderson 2001; Pollack, Erff and Ashkenazi 2001; Fulda, Wick, Weller and Debatin, 2002; LeBlanc, Lawrence, Varfolomeev, Totpal, Morlan, Schow, Fong, Schwall, Sinicropi and Ashkenazi 2002). Unlike other apoptosis-inducing TNF family members, soluble TRAIL appears to be inactive against normal healthy tissue (Ashkenazi, et al. 1999), and reports in which TRAIL induces apoptosis in normal cells could be attributed to the specific preparations of TRAIL used in the experiments (Lawrence, Shahrokh, Marsters, Achilles, Shih, Mounho, Hillan, Totpal, DeForge, Schow, Hooley, Sherwood, Pai, Leung, Khan, Gliniak, Bussiere, Smith, Strom, Kelley, Fox, Thomas and Ashkenazi 2001).

Tumor cells may have an impaired apoptotic response to TRAIL due to resistance mechanism(s) occurring at different points along the TRAIL signaling pathway (Zhang and Fang 2004). Despite mechanism(s) to overcome TRAIL resistance remain largely unclear, a prolonged exposure to the drug or very high doses of TRAIL might be required to overcome resistance (Johnston, Kabore, Strutinsky, Hu, Paul, Kropp, Kuschak, Begleiter and Gibson 2003; Mouzakiti and Packham 2003; Mathas, Lietz, Anagnostopoulos, Hummel, Wiesner, Janz, Jundt, Hirsch, Johrens-Leder, Vornlocher, Bommert, Stein and Dorken 2004; Hasegawa, Yamada, Harasawa, Tsuji, Murata, Sugahara, Tsuruda, Ikeda, Imaizumi, Tomonaga, Masuda, Takasu and Kamihira 2005). However, because of TRAIL's short half-life in plasma (Ashkenazi, et al. 1999) and rapid elimination (Walczak, et al. 1999), achieving prolonged exposure at high concentrations is difficult. Despite in vivo studies using a trimerized (Walczak, et al. 1999) or a non-tagged (Ashkenazi, et al. 1999; Hao, Song, Hsi, Lewis, Song, Petruk, Tyrrell and Kneteman 2004) form of TRAIL have demonstrated a good toxicity profile of the molecule, organ toxicity might occur when using high doses of soluble TRAIL. In experimental anticancer treatments, the response to TRAIL was significantly increased upon co-administration of DNA-damaging chemotherapeutic drugs due to up-regulation of TRAIL-R1 and/or TRAIL-R2 (Wen, Ramadevi, Nguyen, Perkins, Worthington and Bhalla 2000; LeBlanc and Ashkenazi 2003). In addition, irradiation up-regulates TRAIL-R2 receptor expression, and combining irradiation with TRAIL treatment has an additive or synergistic effect (Chinnaiyan, Prasad, Shankar, Hamstra, Shanaiah, Chenevert, Ross and Rehemtulla 2000). Alternatively, up-regulation of TRAIL-R1 or TRAIL-R2 using small molecules, such as the proteasome inhibitor bortezomib (Johnson, Stone, Nikrad, Yeh, Zong, Thompson, Nesterov and Kraft 2003) or inhibitors of histone deacetylase (Nakata, Yoshida, Horinaka, Shiraishi, Wakada and Sakai 2004) might allow to overcome TRAIL resistance.

Gene therapy approaches are currently being developed to specifically target tumor cells and overcome the limitations inherent to TRAIL death receptor targeting, i.e., pharmacokinetic, toxicity profile, pattern of receptor expression, tumor cell resistance. TRAIL-encoding adenoviruses (Ad-TRAIL) might be an alternative for systemic TRAIL delivery, allowing better tumor cell targeting and increased tumoricidal activity (Griffith, Anderson, Davidson, Williams and Ratliff 2000; Griffith and Broghammer 2001; Kagawa, He, Gu, Koch, Rha, Roth, Curley, Stephens and Fang 2001; Lee, Hampl, Albert and Fine 2002; Armeanu, Lauer, Smirnow, Schenk, Weiss, Gregor and Bitzer 2003). However, optimal adenovector gene therapy requires efficient infection of target tumor cells and avoidance of immune clearance (Harrington, Alvarez-Vallina, Crittenden, Gough, Chong, Diaz, Vassaux, Lemoine and Vile 2002). Additionally, safety and toxicity issues of systemic vector administration necessitate intratumoral injection of TRAIL-encoding adenovectors, resulting in local antitumor activity, but with limited value for treating disseminated tumors.

In order to optimize the use of TRAIL-encoding adenovectors, we recently explored a cell therapy approach using a replication-deficient Ad-TRAIL (Griffith. et al. 2000) encoding a full-length membrane-bound TRAIL (mTRAIL) to transduce $CD34^+$ cells ($CD34$-$TRAIL^+$) (Carlo-Stella, Lavazza, Di Nicola, Cleris, Longoni, Milanesi, Magni, Morelli, Gloghini, Carbone and Gianni 2006). Gene-modified $CD34^+$ cells indeed represent optimal vehicles of antitumor molecules. In fact, they can migrate from the bloodstream into tumor tissues due to the expression of adhesion receptors that specifically interact with counter-receptors expressed by tumor endothelial cells (Verfaillie 1998; Kaplan, Riba, Zacharoulis, Bramley, Vincent, Costa, MacDonald, Jin, Shido, Kerns, Zhu, Hicklin, Wu, Port, Altorki, Port, Ruggero, Shmelkov, Jensen, Rafii and Lyden 2005; Burger and Kipps 2006). Additionally, up-regulation of inflammatory chemo-attractants in the tumor micro-environment provides a permissive environment that allows for homing of systemically delivered $CD34$-$TRAIL^+$ cells and efficient tumor targeting (Jin, Aiyer, Su, Borgstrom, Stupack, Friedlander and Varner 2006). Finally, upon adenovector transduction, $CD34^+$ cells express high levels of the transgene for only a few days (Bregni, Shammah, Malaffo,

Di Nicola, Milanesi, Magni, Matteucci, Ravagnani, Jordan, Siena and Gianni 1998). This transient expression, partly related to dilution of the intracellular adenoviral episome through multiple cell divisions, enhances its safety.

8.2 Adenoviral Transduction of CD34+ Cells

Ad-TRAIL-transduced CD34+ cells using a multiplicity of infection (MOI) of 500 expressed high levels of mTRAIL, with a mean (±SD) transduction efficiency of 86 ± 10% (range 70%–96%, n = 10) and a cell viability ≥85%. Lack of TRAIL-R1 and TRAIL-R2 expression on CD34+ cells prevented any activation of the apoptotic signaling by CD34-TRAIL+ cells even following transduction with an MOI as high as 5,000. Similarly, CD34+ cells were unaffected by exposure to soluble TRAIL (100 ng/ml, 48 hours), further demonstrating that high levels of mTRAIL expression do not affect CD34+ cell viability. Western blot analysis of CD34-TRAIL+ cells revealed the presence of 32- and 55-kDa proteins, which are the expected products for full-length monomer and dimer TRAIL, respectively. Time-course analysis of transgene expression revealed that CD34-TRAIL+ cells continued to express significant levels of mTRAIL for at least 96 hours after transduction.

8.3 In Vitro Activity of CD34-TRAIL+ Cells

Upon in vitro co-culture, CD34-TRAIL+ cells exhibited a potent killing activity on a variety of tumor cell types, including lymphoma, multiple myeloma, as well as epithelial cancers. Additionally, mTRAIL-armed cells were highly cytotoxic against tumor cells resistant to soluble TRAIL (Carlo-Stella, et al. 2006). Thus, the membrane-bound form of the ligand is capable of triggering apoptosis more efficiently than soluble TRAIL and overcoming tumor resistance to the soluble ligand. This peculiar functional property of mTRAIL might be due to a differential activation of TRAIL-R1 and TRAIL-R2 by soluble and membrane TRAIL (Muhlenbeck, Schneider, Bodmer, Schwenzer, Hauser, Schubert, Scheurich, Moosmayer, Tschopp and Wajant 2000; Wajant, Moosmayer, Wuest, Bartke, Gerlach, Schonherr, Peters, Scheurich and Pfizenmaier 2001).

To investigate whether cell death triggered by mTRAIL involved the same biochemical signalling pathway of soluble TRAIL, we examined caspase activation and PARP cleavage. Co-culture of the TRAIL-sensitive KMS-11 cell line with CD34-TRAIL+ cells resulted in caspase-3, -8, and -9 activation, and in PARP cleavage, i.e., the same pattern of biochemical events achieved by incubating KMS-11 cells with soluble TRAIL. Caspase activation and PARP cleavage were also observed by exposing the TRAIL-resistant JVM-2 cell line to mTRAIL-armed cells, but not to soluble TRAIL. No caspase activation or PARP cleavage was detected by co-culturing KMS-11 or JVM-2 cells with mock-transduced CD34+

cells, thus excluding non-specific cell toxic effects caused by the adenovector. Western blot analysis of CD34-TRAIL⁺ cells showed no caspase activation or PARP cleavage, thus ruling out the possibility that transduced CD34⁺ cells, rather than tumor cells, might have contributed to caspase activation. Overall, these data strongly suggest that the antitumor activity of CD34-TRAIL⁺ cells effectively depends on TRAIL receptor ligation by mTRAIL-expressing CD34⁺ cells.

8.4 In Vivo Antitumor Activity of CD34-TRAIL⁺ Cells

In addition to in vitro cell killing effects, CD34-TRAIL⁺ cells showed potent in vivo tumoricidal activity in non-obese diabetic/severe combined immuno-deficient (NOD/SCID) mice xenografted with either the TRAIL-sensitive KMS-11 or the TRAIL-resistant JVM-2 cell lines. Repeated injections of mTRAIL-armed cells to NOD/SCID mice bearing an advanced-stage KMS-11 xenograft resulted in a significant increase of median survival over controls (83 vs 55 days, $P \leq 0.0001$), with 28 percent of NOD/SCID mice being alive and disease-free at the end of the 150-day observation period (Fig. 8-1a). Additionally, mTRAIL-armed cells induced a significantly better survival as compared to soluble TRAIL (83 vs. 66 days, $P \leq 0.0001$) (Fig. 8-1a). To investigate whether CD34-TRAIL⁺ cells also could overcome soluble TRAIL-resistance in vivo, mTRAIL cells were tested in mice xenografted with the soluble TRAIL-resistant JVM-2 cell line. As compared with controls, four weekly injections of CD34-TRAIL⁺

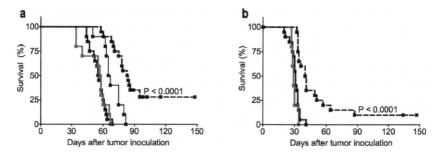

Fig. 8-1 (a) Survival curves of NOD/SCID mice xenografted with the TRAIL-sensitive KMS-11 cell line (0.5×10^6 cells/mouse) and treated at an advanced-stage of disease. Mice received four weekly intravenous injections of PBS (n = 20, black line), CD34-mock cells (n = 20, gray line), CD34-TRAIL⁺ cells (n = 20, dotted line), and soluble TRAIL (n = 10, thin line). Weekly administrations of mock- or Ad-TRAIL–transduced CD34⁺ cells (1×10^6 cells/mouse) were started 14 days after tumor injection. Mice treated with recombinant soluble TRAIL received four weekly subcutaneous injections (50 µg/mouse/injection) starting 14 days after tumor inoculation. (b) Survival curves of NOD/SCID mice xenografted with the TRAIL-resistant JVM-2 cell line (1×10^6 cells/mouse). Mice received four weekly intravenous injections of PBS (n = 20, black line), CD34-mock cells (n = 10, gray line), and CD34-TRAIL⁺ cells (n = 20, dotted line). Weekly administrations of mock- or Ad-TRAIL–transduced CD34⁺ cells (1×10^6 cells/mouse) were started on day 1 after tumor inoculation

cells resulted in a significant prolongation of the median survival times of mice xenografted with JVM-2 (40 vs. 31 days, $P \leq 0.0001$) cells (Fig. 8-1b).

8.5 Toxicity of CD34-TRAIL$^+$ Cells

The issue of TRAIL-induced liver toxicity was investigated in NOD/SCID mice receiving a single injection of Ad-TRAIL-transduced cells (5×10^6 cells/mouse). Weekly monitoring for liver enzyme activity showed no evidence of liver toxicity, as compared to controls. This outcome was not the result of a lack of human mTRAIL activity on mouse tissues, as suggested by the previously reported high sequence homology of murine and human TRAIL-R2 (Wu, Burns, Zhan, Alnemri and El-Deiry 1999), the constitutive liver expression of TRAIL-R2 (Zheng, Wang, Tsabary and Chen 2004), and the cross-species activity of human and murine TRAIL proteins (Walczak, et al. 1999). Furthermore, human mTRAIL activity in mice was confirmed by injecting animals with the Ad-TRAIL vector, which localizes to the liver (Lieber, He, Meuse, Schowalter, Kirillova, Winther and Kay 1997). This procedure resulted in an early appearing and long-lasting liver toxicity. Additional studies were carried out using an adenovector encoding the murine TRAIL (kindly provided by Dr. A. Gambotto, Vector Core Facility, University of Pittsburgh, Pittsburgh, PA, USA). As expected, the biological activity of the murine molecule was higher, and the intravenous injection of 1×10^9 pfu/mouse induced an average AST and ALT increase of 20- and 65-fold, respectively. Conversely, mice receiving 5×10^6 CD34$^+$ cells transduced with the same vector, and expressing high levels of the murine TRAIL failed to show evidence of liver toxicity. The latter experiments, which more closely mimic a clinical scenario of somatic cell therapy, suggest that cell-mediated TRAIL administration does not result in liver toxicity, the greatest toxicity concern for systemically administered TNF family members.

8.6 Histologic Analysis

To gain insight into the antitumor mechanism(s) of mTRAIL-armed cells, we addressed the issue of tumor homing of transduced cells, and analyzed the degree and distribution of tumor cell death, and the status of tumor vasculature. Tumor homing of transduced cells was analyzed in NOD/SCID mice bearing subcutaneous tumor nodules. Following a single intravenous injection of CD34-TRAIL$^+$ cells, nodules were excised at different time-points and immuno-stained with the anti-human CD45 antibody. CD34-TRAIL$^+$ cells were detected in the tumors as early as 30 minutes following injection, peaked after 24 hours and persisted up to 48 hours post-treatment. At peak, the percentage of CD45$^+$ cells per tumor section ranged from 0.2 to 0.4 percent, as evaluated using the NIH imaging software ImageJ.

Consistent with the early tumor homing of transduced cells, hematoxylin and eosin staining revealed that injection of CD34-TRAIL+ cells, but not mock-transduced CD34+ cells, induced tumor hemorrhage that could be detected as early as 0.5 hour following treatment, and necrosis which was detectable five hours following infusion (Fig. 8-2). Hemorrhagic and necrotic findings initially detected in the peripheral area of the tumor extended over time to the inner portion of the nodule. TUNEL staining revealed the presence of apoptotic cells, detectable as early as 0.5 hour post-treatment and increasing in a time-dependent manner up to 120 hour post-injection, when a 21-fold increase of the apoptotic index was detected (Fig. 8-2). Because the early appearing apoptotic and necrotic areas observed in mice receiving CD34-TRAIL+ cells were distributed along the tumor vasculature, we analyzed tumor blood vessels in mice treated with mock-transduced CD34+ cells or CD34-TRAIL+ cells by staining endothelial cells with a monoclonal antibody directed against the vascular endothelial growth factor receptor-3 (VEGFR3). As early as 0.5 hour following injection of CD34-TRAIL+ cells, we observed signs of vascular damage leading to a progressive disintegration of the vascular bed, which colocalized with apoptotic and necrotic phenomena.

These results suggest that tumor endothelial cells represent an early target of CD34-TRAIL$^+$ cells. Staining with an anti-TRAIL-R2 antibody showed that this receptor is widely expressed on tumor endothelial cells, suggesting that binding of CD34-TRAIL$^+$ cells to TRAIL-R2 might represent the first step of a complex cascade of events leading to tumor cell death and involving both direct tumor cell killing due to apoptosis and indirect tumor cell killing due to vascular-disrupting mechanisms.

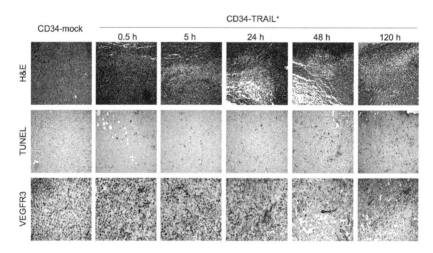

Fig. 8-2 Hematoxylin and eosin (H&E) (magnification, x20), TUNEL (magnification, x20) and anti-mouse VEGFR-3 (magnification, x40) staining of sections from tumor nodules growing in mice injected subcutaneously with KMS-11 cell line. Tumor nodules were excised at different time-points following injections of mock-transduced CD34$^+$ cells or CD34-TRAIL$^+$ cells (5 × 10^6/mouse). Injection of mTRAIL-expressing cells was associated with intratumor hemorrhage, apoptosis, necrosis, and vascular damage

8.7 Conclusions

Since death receptor activation can instruct malignant cells to undergo apoptosis independent of p53, targeting death receptors with TRAIL-targeting therapeutics is a rational therapeutic strategy against cancer. TRAIL-targeting therapeutics possess considerable and specific antitumor activity, both when used alone as well as in combination with nonspecific cytotoxic agents, radiation, and other target-based therapeutics. However, optimization of death receptor targeting requires overcoming limitations intrinsic to soluble TRAIL, i.e., pharmacokinetic, toxicity profile, pattern of receptor expression, tumor cell resistance.

Here, we describe a novel anticancer cell therapy approach, using $CD34^+$ cells for systemic delivery of the membrane-bound form of TRAIL to tumors. In vitro results show that $CD34^+$ cells were easily and reproducibly transduced by an adenovector containing full-length TRAIL, expressed the transgene at high level, and were totally unaffected by TRAIL expression on their surface. In vitro, CD34-$TRAIL^+$ cells exhibited high killing activity on a variety of tumor cell types and were highly cytotoxic against TRAIL-resistant tumor cells. In vivo, CD34-$TRAIL^+$ cells showed potent tumoricidal activity. Repeated intravenous dosing with mTRAIL-armed cells clearly resulted in elimination of tumor cells with survival being significantly prolonged in tumor-bearing mice. Interestingly, in mice with an advanced-stage KMS-11 xenograft, the antitumor activity of mTRAIL-armed cells was higher than that induced by soluble TRAIL. A therapeutic activity of mTRAIL-expressing cells could also be detected in mice xenografted with soluble TRAIL-resistant cell lines, suggesting that transduced cells might at least partially overcome resistance to soluble TRAIL.

$CD34^+$ cells were selected as vehicles for mTRAIL delivery due to their migratory properties as well as capacity of specifically interacting with endothelial cells through receptor-mediated mechanisms (Verfaillie 1998; Lapidot, Dar and Kollet 2005). Due to inflammatory characteristics of tumor micro-environment, systemically delivered $CD34^+$ cells can efficiently home in the tumor by rolling on tumor vasculature, adhering to it and finally extravasating in the tumor (Jin, et al. 2006). In our study, genetically modified $CD34^+$ cells exhibited a tumor-homing capacity similar to the bone marrow homing capacity of $CD34^+$ cells injected in non-irradiated recipients (De Palma, Venneri, Roca and Naldini 2003; Quesenberry, Colvin and Abedi 2005). Tumor-homing of CD34-$TRAIL^+$ cells was followed by a direct tumor cell killing effect due to apoptosis. However, the expression of TRAIL-R2 on tumor endothelial cells, as well as the characteristic perivascular distribution of tumor cell death, suggests that CD34-$TRAIL^+$ cells also act through an indirect tumor cell killing mechanism due to vascular disruption. Thus, binding of CD34-$TRAIL^+$ cells to TRAIL-R2 might represent an early step of a complex cascade of events resulting in extensive vascular damage and hemorrhagic necrosis of the tumor.

Genetically-modified neural or mesenchymal stem cells have been demonstrated to exert an efficient antitumor activity following intratumor or systemic injection in xenograft models of glioma (Ehtesham, Kabos, Gutierrez, Chung, Griffith, Black and Yu 2002; Ehtesham, Kabos, Kabosova, Neuman, Black and Yu 2002; Lee,

Elkahloun, Messina, Ferrari, Xi, Smith, Cooper, Albert and Fine 2003; Nakamizo, Marini, Amano, Khan, Studeny, Gumin, Chen, Hentschel, Vecil, Dembinski, Andreeff and Lang 2005), melanoma (Studeny, Marini, Champlin, Zompetta, Fidler and Andreeff 2002), and breast cancer (Studeny, Marini, Dembinski, Zompetta, Cabreira-Hansen, Bekele, Champlin and Andreeff 2004). Our data demonstrate for the first time that targeting of disseminated lympho-hematopoietic tumors can also be achieved by intravenous injection of CD34$^+$ cells. These cells are already widely used in the clinical setting and might easily become a therapeutically applicable approach. Investigating whether CD34$^+$ cells exert a better therapeutic activity than neural or mesenchymal cells remains an issue to be addressed by future studies. Comparing the efficacy of various cell carriers will clarify whether distinct tumor types or tumor localizations (i.e. lung metastasis vs. bone metastasis) might preferentially benefit from a specific cell type. Clinical translation of gene therapy approaches using adenovector-transduced cells for delivery of membrane-bound TRAIL represent a challenging strategy that might achieve systemic tumor targeting and increased intra-tumor delivery of the therapeutic agent.

Acknowledgments Supported in part by grants from Ministero dell'Università e della Ricerca (MUR, Rome, Italy), Ministero della Salute (Rome, Italy), and Michelangelo Foundation for Advances in Cancer Research and Treatment (Milano, Italy).

References

Almasan, A., and Ashkenazi, A. (2003) Apo2L/TRAIL: apoptosis signaling, biology, and potential for cancer therapy. Cytokine Growth Factor Rev. 14, 337–348.

Armeanu, S., Lauer, U. M., Smirnow, I., Schenk, M., Weiss, T. S., Gregor, M., and Bitzer, M. (2003) Adenoviral gene transfer of tumor necrosis factor-related apoptosis-inducing ligand overcomes an impaired response of hepatoma cells, but causes severe apoptosis in primary human hepatocytes. Cancer Res. 63, 2369–2372.

Ashkenazi, A. (2002) Targeting death and decoy receptors of the tumor-necrosis factor superfamily. Nat Rev Cancer. 2, 420–430.

Ashkenazi, A., Pai, R. C., Fong, S., Leung, S., Lawrence, D. A., Marsters, S. A., Blackie, C., Chang, L., Mcmurtrey, A. E., Hebert, A., Deforge, L., Koumenis, I. L., Lewis, D., Harris, L., Bussiere, J., Koeppen, H., Shahrokh, Z., and Schwall, R. H. (1999) Safety and antitumor activity of recombinant soluble Apo2 ligand. J. Clin. Invest. 104, 155–162.

Bregni, M., Shammah, S., Malaffo, F., Di Nicola, M., Milanesi, M., Magni, M., Matteucci, P., Ravagnani, F., Jordan, C.T., Siena, S., and Gianni, A. M. (1998) Adenovirus vectors for gene transduction into mobilized blood CD34+ cells. Gene Ther. 5, 465–472.

Burger, J. A., and Kipps, T. J. (2006) CXCR4: a key receptor in the crosstalk between tumor cells and their micro-environment. Blood. 107, 1761–1767.

Carlo-Stella, C., Lavazza, C., Di Nicola, M., Cleris, L., Longoni, P., Milanesi, M., Magni, M., Morelli, D., Gloghini, A., Carbone, A., and Gianni, A. M. (2006) Antitumor Activity of Human CD34(+) Cells Expressing Membrane-Bound Tumor Necrosis Factor-Related Apoptosis-Inducing Ligand. Hum Gene Ther. 17, 1225–1240.

Chinnaiyan, A. M., Prasad, U., Shankar, S., Hamstra, D. A., Shanaiah, M., Chenevert, T. L., Ross, B. D., and Rehemtulla, A. (2000) Combined effect of tumor necrosis factor-related apoptosis-inducing ligand and ionizing radiation in breast cancer therapy. Proc Natl Acad Sci U S A. 97, 1754–1759.

De Palma, M., Venneri, M. A., Roca, C., and Naldini, L. (2003) Targeting exogenous genes to tumor angiogenesis by transplantation of genetically modified hematopoietic stem cells. Nat Med. 9, 789–795.

Degli-Esposti, M. A., Dougall, W. C., Smolak, P. J., Waugh, J. Y., Smith, C. A., and Goodwin, R. G. (1997) The novel receptor TRAIL-R4 induces NF-kappaB and protects against TRAIL-mediated apoptosis, yet retains an incomplete death domain. Immunity. 7, 813–820.

Ehtesham, M., Kabos, P., Gutierrez, M. A. R., Chung, N. H. C., Griffith, T. S., Black, K. L., and Yu, J. S. (2002) Induction of Glioblastoma Apoptosis Using Neural Stem Cell-mediated Delivery of Tumor Necrosis Factor-related Apoptosis-inducing Ligand. Cancer Res. 62, 7170–7174.

Ehtesham, M., Kabos, P., Kabosova, A., Neuman, T., Black, K. L., and Yu, J. S. (2002) The use of interleukin 12-secreting neural stem cells for the treatment of intracranial glioma. Cancer Res. 62, 5657–5663.

Emery, J. G., Mcdonnell, P., Burke, M. B., Deen, K. C., Lyn, S., Silverman, C., Dul, E., Appelbaum, E. R., Eichman, C., Diprinzio, R., Dodds, R. A., James, I. E., Rosenberg, M., Lee, J. C., and Young, P. R. (1998) Osteoprotegerin is a receptor for the cytotoxic ligand TRAIL. J Biol Chem. 273, 14363–14367.

Fulda, S., Wick, W., Weller, M., and Debatin, K. -M. (2002) Smac agonists sensitize for Apo2L/TRAIL- or anticancer drug-induced apoptosis and induce regression of malignant glioma in vivo. Nature Medicine. 8, 808–815.

Griffith, T. S., Anderson, R. D., Davidson, B. L., Williams, R. D., and Ratliff, T. L. (2000) Adenoviral-mediated transfer of the TNF-related apoptosis-inducing ligand/Apo-2 ligand gene induces tumor cell apoptosis. J Immunol. 165, 2886–2894.

Griffith, T. S., and Broghammer, E. L. (2001) Suppression of tumor growth following intralesional therapy with TRAIL recombinant adenovirus. Mol Ther. 4, 257–266.

Hao, C., Song, J. H., Hsi, B., Lewis, J., Song, D. K., Petruk, K. C., Tyrrell, D. L., and Kneteman, N. M. (2004) TRAIL inhibits tumor growth but is nontoxic to human hepatocytes in chimeric mice. Cancer Res. 64, 8502–8506.

Harrington, K., Alvarez-Vallina, L., Crittenden, M., Gough, M., Chong, H., Diaz, R. M., Vassaux, G., Lemoine, N., and Vile, R. (2002) Cells as vehicles for cancer gene therapy: the missing link between targeted vectors and systemic delivery? Hum Gene Ther. 13, 1263–1280.

Hasegawa, H., Yamada, Y., Harasawa, H., Tsuji, T., Murata, K., Sugahara, K., Tsuruda, K., Ikeda, S., Imaizumi, Y., Tomonaga, M., Masuda, M., Takasu, N., and Kamihira, S. (2005) Sensitivity of adult T cell leukaemia lymphoma cells to tumor necrosis factor-related apoptosis-inducing ligand. Br J Haematol. 128, 253–265.

Jin, H., Aiyer, A., Su, J., Borgstrom, P., Stupack, D., Friedlander, M., and Varner, J. (2006) A homing mechanism for bone marrow-derived progenitor cell recruitment to the neovasculature. J Clin Invest. 116, 652–662.

Johnson, T. R., Stone, K., Nikrad, M., Yeh, T., Zong, W. X., Thompson, C. B., Nesterov, A., and Kraft, A. S. (2003) The proteasome inhibitor PS-341 overcomes TRAIL resistance in Bax and caspase 9-negative or Bcl-xL over-expressing cells. Oncogene. 22, 4953–4963.

Johnston, J. B., Kabore, A. F., Strutinsky, J., Hu, X., Paul, J. T., Kropp, D. M., Kuschak, B., Begleiter, A., and Gibson, S. B. (2003) Role of the TRAIL/APO2-L death receptors in chlorambucil- and fludarabine-induced apoptosis in chronic lymphocytic leukemia. Oncogene. 22, 8356–8369.

Kagawa, S., He, C., Gu, J., Koch, P., Rha, S. -J., Roth, J. A., Curley, S. A., Stephens, L. C., and Fang, B. (2001) Antitumor activity and bystander effects of the tumor necrosis factor-related apoptosis-inducing ligand (TRAIL) gene. Cancer Res. 61, 3330–3338.

Kaplan, R. N., Riba, R. D., Zacharoulis, S., Bramley, A. H., Vincent, L., Costa, C., Macdonald, D. D., Jin, D. K., Shido, K., Kerns, S. A., Zhu, Z., Hicklin, D., Wu, Y., Port, J. L., Altorki, N., Port, E. R., Ruggero, D., Shmelkov, S. V., Jensen, K. K., Rafii, S., and Lyden, D. (2005) VEGFR1-positive haematopoietic bone marrow progenitors initiate the pre-metastatic niche. Nature. 438, 820–827.

Laconi, E., Pani, P., and Farber, E. (2000) The resistance phenotype in the development and treatment of cancer. Lancet Oncol. 1, 235–241.

Lapidot, T., Dar, A., and Kollet, O. (2005) How do stem cells find their way home? Blood. 106, 1901–1910.

Lawrence, D., Shahrokh, Z., Marsters, S., Achilles, K., Shih, D., Mounho, B., Hillan, K., Totpal, K., Deforge, L., Schow, P., Hooley, J., Sherwood, S., Pai, R., Leung, S., Khan, L., Gliniak, B., Bussiere, J., Smith, C. A., Strom, S. S., Kelley, S., Fox, J. A., Thomas, D., and Ashkenazi, A. (2001) Differential hepatocyte toxicity of recombinant Apo2L/TRAIL versions. Nat Med. 7, 383–385.

Leblanc, H., Lawrence, D., Varfolomeev, E., Totpal, K., Morlan, J., Schow, P., Fong, S., Schwall, R., Sinicropi, D., and Ashkenazi, A. (2002) Tumor-cell resistance to death receptor–induced apoptosis through mutational inactivation of the proapoptotic Bcl-2 homolog Bax. Nat Med. 8, 274–281.

Leblanc, H. N., and Ashkenazi, A. (2003) Apo2L/TRAIL and its death and decoy receptors. Cell Death Differ. 10, 66–75.

Lee, J., Elkahloun, A. G., Messina, S. A., Ferrari, N., Xi, D., Smith, C. L., Cooper, R., Jr., Albert, P. S., and Fine, H. A. (2003) Cellular and genetic characterization of human adult bone marrow-derived neural stem-like cells: a potential antiglioma cellular vector. Cancer Res. 63, 8877–8889.

Lee, J., Hampl, M., Albert, P., and Fine, H. A. (2002) Antitumor activity and prolonged expression from a TRAIL-expressing adenoviral vector. Neoplasia (New York, N.Y.). 4, 312–323.

Lieber, A., He, C. Y., Meuse, L., Schowalter, D., Kirillova, I., Winther, B., and Kay, M. A. (1997) The role of Kupffer cell activation and viral gene expression in early liver toxicity after infusion of recombinant adenovirus vectors. J Virol. 71, 8798–8807.

Mariani, S. M., Matiba, B., Armandola, E. A., and Krammer, P. H. (1997) Interleukin 1 beta-converting enzyme related proteases/caspases are involved in TRAIL-induced apoptosis of myeloma and leukemia cells. J Cell Biol. 137, 221–229.

Mathas, S., Lietz, A., Anagnostopoulos, I., Hummel, F., Wiesner, B., Janz, M., Jundt, F., Hirsch, B., Johrens-Leder, K., Vornlocher, H. P., Bommert, K., Stein, H., and Dorken, B. (2004) c-FLIP mediates resistance of Hodgkin/Reed-Sternberg cells to death receptor-induced apoptosis. J Exp Med. 199, 1041–1052.

Mitsiades, C. S., Treon, S. P., Mitsiades, N., Shima, Y., Richardson, P., Schlossman, R., Hideshima, T., and Anderson, K. C. (2001) TRAIL/Apo2L ligand selectively induces apoptosis and overcomes drug resistance in multiple myeloma: therapeutic applications. Blood. 98, 795–804.

Mouzakiti, A., and Packham, G. (2003) Regulation of tumor necrosis factor-related apoptosis-inducing ligand (TRAIL)-induced apoptosis in Burkitt's lymphoma cell lines. Br J Haematol. 122, 61–69.

Muhlenbeck, F., Schneider, P., Bodmer, J. L., Schwenzer, R., Hauser, A., Schubert, G., Scheurich, P., Moosmayer, D., Tschopp, J., and Wajant, H. (2000) The tumor necrosis factor-related apoptosis-inducing ligand receptors TRAIL-R1 and TRAIL-R2 have distinct cross-linking requirements for initiation of apoptosis and are non-redundant in JNK activation. J Biol Chem. 275, 32208–32213.

Nakamizo, A., Marini, F., Amano, T., Khan, A., Studeny, M., Gumin, J., Chen, J., Hentschel, S., Vecil, G., Dembinski, J., Andreeff, M., and Lang, F. F. (2005) Human bone marrow-derived mesenchymal stem cells in the treatment of gliomas. Cancer Res. 65, 3307–3318.

Nakata, S., Yoshida, T., Horinaka, M., Shiraishi, T., Wakada, M., and Sakai, T. (2004) Histone deacetylase inhibitors upregulate death receptor 5/TRAIL-R2 and sensitize apoptosis induced by TRAIL/APO2-L in human malignant tumor cells. Oncogene. 23, 6261–6271.

Pan, G., Ni, J., Wei, Y. F., Yu, G., Gentz, R., and Dixit, V. M. (1997) An antagonist decoy receptor and a death domain-containing receptor for TRAIL. Science. 277, 815–818.

Pan, G., O'Rourke, K., Chinnaiyan, A. M., Gentz, R., Ebner, R., Ni, J., and Dixit, V. M. (1997) The receptor for the cytotoxic ligand TRAIL. Science. 276, 111–113.

Pitti, R. M., Marsters, S. A., Ruppert, S., Donahue, C. J., Moore, A., and Ashkenazi, A. (1996) Induction of apoptosis by Apo-2 ligand, a new member of the tumor necrosis factor cytokine family. J Biol Chem. 271, 12687–12690.

Pollack, I. F., Erff, M., and Ashkenazi, A. (2001) Direct stimulation of apoptotic signalling by soluble Apo2L/tumor necrosis factor-related apoptosis-inducing ligand leads to selective killing of glioma cells. Clinical Cancer Research. 7, 1362–1369.

Pommier, Y., Sordet, O., Antony, S., Hayward, R. L., and Kohn, K. W. (2004) Apoptosis defects and chemotherapy resistance: molecular interaction maps and networks. Oncogene. 23, 2934–2949.

Quesenberry, P. J., Colvin, G., and Abedi, M. (2005) Perspective: fundamental and clinical concepts on stem cell homing and engraftment: a journey to niches and beyond. Exp Hematol. 33, 9–19.

Sheridan, J. P., Marsters, S. A., Pitti, R. M., Gurney, A., Skubatch, M., Baldwin, D., Ramakrishnan, L., Gray, C. L., Baker, K., Wood, W. I., Goddard, A. D., Godowski, P., and Ashkenazi, A. (1997) Control of TRAIL-induced apoptosis by a family of signalling and decoy receptors. Science. 277, 818–821.

Studeny, M., Marini, F. C., Champlin, R. E., Zompetta, C., Fidler, I. J., and Andreeff, M. (2002) Bone marrow-derived mesenchymal stem cells as vehicles for interferon-beta delivery into tumors. Cancer Res. 62, 3603–3608.

Studeny, M., Marini, F. C., Dembinski, J. L., Zompetta, C., Cabreira-Hansen, M., Bekele, B. N., Champlin, R. E., and Andreeff, M. (2004) Mesenchymal stem cells: potential precursors for tumor stroma and targeted-delivery vehicles for anticancer agents. J Natl Cancer Inst. 96, 1593–1603.

Verfaillie, C. M. (1998) Adhesion receptors as regulators of the hematopoietic process. Blood. 92, 2609–2612.

Wajant, H. (2003) Death receptors. Essays Biochem. 39, 53–71.

Wajant, H., Moosmayer, D., Wuest, T., Bartke, T., Gerlach, E., Schonherr, U., Peters, N., Scheurich, P., and Pfizenmaier, K. (2001) Differential activation of TRAIL-R1 and -2 by soluble and membrane TRAIL allows selective surface antigen-directed activation of TRAIL-R2 by a soluble TRAIL derivative. Oncogene. 20, 4101–4106.

Walczak, H., Degli-Esposti, M. A., Johnson, R. S., Smolak, P. J., Waugh, J. Y., Boiani, N., Timour, M. S., Gerhart, M. J., Schooley, K. A., Smith, C. A., Goodwin, R. G., and Rauch, C. T. (1997) TRAIL-R2: a novel apoptosis-mediating receptor for TRAIL. Embo J. 16, 5386–5397.

Walczak, H., Miller, R. E., Ariail, K., Gliniak, B., Griffith, T. S., Kubin, M., Chin, W., Jones, J., Woodward, A., Le, T., Smith, C., Smolak, P., Goodwin, R. G., Rauch, C. T., Schuh, J. C., and Lynch, D. H. (1999) Tumoricidal activity of tumor necrosis factor-related apoptosis-inducing ligand in vivo. Nat Med. 5, 157–163.

Wang, S., and El-Deiry, W. S. (2003) TRAIL and apoptosis induction by TNF-family death receptors. Oncogene. 22, 8628–8633.

Waxman, D. J., and Schwartz, P. S. (2003) Harnessing apoptosis for improved anticancer gene therapy. Cancer Res. 63, 8563–8572.

Wen, J., Ramadevi, N., Nguyen, D., Perkins, C., Worthington, E., and Bhalla, K. (2000) Antileukemic drugs increase death receptor 5 levels and enhance Apo-2L-induced apoptosis of human acute leukemia cells. Blood. 96, 3900–3906.

Wiley, S. R., Schooley, K., Smolak, P. J., Din, W. S., Huang, C. P., Nicholl, J. K., Sutherland, G. R., Smith, T. D., Rauch, C., and Smith, C. A. (1995) Identification and characterization of a new member of the TNF family that induces apoptosis. Immunity. 3, 673–682.

Wu, G. S., Burns, T. F., Zhan, Y., Alnemri, E. S., and El-Deiry, W. S. (1999) Molecular cloning and functional analysis of the mouse homologue of the KILLER/DR5 tumor necrosis factor-related apoptosis-inducing ligand (TRAIL) death receptor. Cancer Res. 59, 2770–2775.

Zhang, L., and Fang, B. (2004) Mechanisms of resistance to TRAIL-induced apoptosis in cancer. Cancer Gene Ther.

Zheng, S. J., Wang, P., Tsabary, G., and Chen, Y. H. (2004) Critical roles of TRAIL in hepatic cell death and hepatic inflammation. J Clin Invest. 113, 58–64.

9
Linking Inflammation Reactions to Cancer: Novel Targets for Therapeutic Strategies

Alberto Mantovani[1,2], Federica Marchesi[1], Chiara Porta[1], Paola Allavena[1], and Antonio Sica[1]

9.1 Introduction

An inflammatory component is present in the microenvironment of most neoplastic tissues, including those not causally related to an obvious inflammatory process. Epidemiological studies have revealed that chronic inflammation predisposes to different forms of cancer and that usage of non-steroidal, anti-inflammatory agents is associated with protection against various tumors. The infiltration of white blood cells, the presence of polypeptide messengers of inflammation (cytokines and chemokines), the occurrence of tissue remodeling and angiogenesis, represent hallmarks of cancer-associated inflammation.

In the late 1970s, it was found that a major leukocyte population present in tumors – the so-called tumor-associated macrophage (TAM) – promotes tumor growth (Balkwill and Mantovani 2001; Mantovani, et al. 2002). Accordingly, in many, but not all, human tumors, a high frequency of infiltrating TAM is associated with poor prognosis. Interestingly, this pathological finding has re-emerged in the post genomic era: genes associated to leukocyte or macrophage infiltration (i.e., CD68) are part of molecular signatures which herald poor prognosis in lymphomas and breast carcinoma (Paik, et al. 2004).

Gene-modified mice, including some with cell-specific targeted gene inactivation, allowed dissection of molecular pathways of inflammation leading to tumor promotion, as well as the initial analysis of the role of distinct elements of the inflammatory process in different steps of tumor progression. TNF, IL-1, the macrophage growth and attractant factor CSF-1, CCL2, a chemokine originally described as a tumor-derived macrophage attractant, the prostaglandin-producing enzyme cyclooxygenase 2, the master inflammatory transcription factor NF-κB, enzymes involved in tissue remodeling, all are essential elements for carcinogenesis and/or for acquisition of a metastatic phenotype in diverse organs including skin,

---

[1] Istituto Clinico Humanitas IRCCS, Via Manzoni 56, 20089 Rozzano (Milan), Italy

[2] Centro di Eccellenza per l'Innovazione Diagnostica e Terapeutica, Institute of Pathology, University of Milan, Italy. Alberto.mantovani@humanitas.it

F. Colotta and A. Mantovani (eds.), *Targeted Therapies in Cancer*.
© Springer 2008

liver, mammary gland and intestine (Coussens and Werb 2002; Voronov, et al. 2003; Greten, et al. 2004; Koehne and Dubois. 2004; Pikarsky et al. 2004; Wyckoff, et al. 2004; Garlanda, et al. 2007, in press).

Here we will review available information on the role of myelomonocytic cells – tumor-associated macrophages (TAMs) in particular – in tumor invasion and metastasis, and discuss potential therapeutic targets.

9.2 Macrophage Polarization

Heterogeneity and plasticity are hallmarks of cells belonging to the monocyte-macrophage lineage (Mantovani, et al. 2002; Gordon 2003; Mantovani, et al. 2004). Lineage-defined populations of mononuclear phagocytes have not been identified, but already the short-lived stage of circulating precursor monocyte subsets characterized by differential expression of the FcγRIII receptor (CD16) or of chemokine receptors (CCR2, CX3CR1 and CCR8) and by different functional properties have been described. Once in tissues macrophages acquire distinct morphological and functional properties directed by the tissue (i.e., the lung alveolar macrophage) and immunological micro-environment.

In response to cytokines and microbial products, mononuclear phagocytes express specialized and polarized functional properties (Gordon 2003; Mantovani, et al. 2004; Ghassabeh, et al. 2006). Mirroring the Th1/Th2 nomenclature, many refer to polarized macrophages as M1 and M2 cells. Classically activated M1 macrophages have long been known to be induced by IFNγ alone or in concert with microbial stimuli (i.e., LPS) or cytokines (i.e., TNF and GM-CSF). IL-4 and IL-13 were subsequently found to be more than simple inhibitors of macrophage activation and to induce an alternative M2 form of macrophage activation (Gordon 2003). M2 is a generic name for various forms of macrophage activation other than the classic M1, including cells exposed to IL-4 or IL-13, immune complexes, IL-10, glucocorticoid, or secosteroid hormones (Mantovani, et al. 2004).

In general, M1 cells have an IL-12^{high}, IL-23^{high}, IL-10^{low} phenotype; are efficient producers of effector molecules (reactive oxygen and nitrogen intermediates) and inflammatory cytokines (IL-1β, TNF, IL-6); participate as inducer and effector cells in polarized Th1 responses; mediate resistance against intracellular parasites and tumors. In contrast, the various forms of M2 macrophages share an IL-12^{low}, IL-23^{low}, IL-10^{high} phenotype with variable capacity to produce inflammatory cytokines depending on the signal utilized. M2 cells generally have high levels of scavenger-, mannose- and galactose-type receptors, and the arginine metabolism is shifted to ornithine and polyamines. Differential regulation of components of the IL-1 system (Dinarello 2005) occurs in polarized macrophages, with low IL-1β and low caspase I, high IL-1ra and high decoy Type II receptor in M2 cells. M1 and the various forms of M2 cells have distinct chemokine and chemokine receptor repertoires (Mantovani, et al. 2004). In general, M2 cells participate in polarized Th2 reactions, promote killing and encapsulation of parasites (Noel, et al, 2004), are present in established tumors and promote progression, tissue repair and remodelling (Wynn, 2004), and have immunoregulatory functions (Mantovani, et al, 2004). Immature myeloid suppressor cells have functional properties and a transcriptional profile related to M2 cells (Biswas, et al. 2006). Moreover, polarization of neutrophil functions has also been reported (Tsuda, et al. 2004).

Profiling techniques and genetic approaches have been tested and shed new light on the M1/M2 paradigm. Transcriptional profiling has offered a comprehensive picture of the genetic programs activated in polarized macrophages, led to the discovery of new polarization-associated genes (i.e., Fizz and YM-1), tested the validity of the paradigm in vivo in selected diseases (Takahashi, et al. 2004; Desnues, et al. 2005; Biswas, et al. 2006), and questioned the generality of some assumptions. For instance, unexpectedly arginase is not expressed prominently in human IL-4-induced M2 cells (Scotton, et al. 2005). M2 cells express high levels of the chitinase-like YM-1. Chitinases represent an anti-parasite strategy conserved in evolution and there is now evidence that acidic mammalian chitinase induced by IL-13 in macrophages is an important mediator of Type II inflammation (Zhu, et al, 2004).

9.3 Macrophage Recruitment at the Tumor Site

Since the first observation by Rudolf Virchow, who noticed the infiltration of leukocytes into malignant tissues and suggested that cancers arise at regions of chronic inflammation, the origin of TAM has been studied in terms of recruitment, survival and proliferation. TAMs derive from circulating monocytes and are recruited at the tumor site by a tumor-derived chemotactic factor for monocytes, originally described by this group (Bottazzi, et al. 1983) and later identified as the chemokine CCL2/MCP-1 (Yoshimura, et al. 1987; Matsushima, et al. 1989) (Fig. 9-1). Evidence supporting a pivotal role of chemokines in the recruitment of monocytes in neoplastic tissues includes: correlation between production and infiltration in murine and human tumors, passive immunization and gene modification (Rollins 1997). In addition, the central role of chemokines in shaping the tumor microenvironment is supported by the observation that tumors are generally characterized

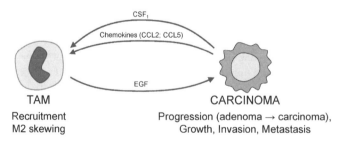

Fig. 9-1 The interplay between carcinoma cells and TAM in cancer

by the constitutive expression of chemokines belonging to the inducible realm (Mantovani 1999). The molecular mechanisms accounting for the constitutive expression of chemokines by cancer cells have been defined only for CXCL1 and involve NF-κB activation by NF-κB-inducing kinase (Yang and Richmond 2001). CCL2 is probably the most frequently found CC chemokine in tumors. Most human carcinomas produce CCL2 and its levels of expression correlate with the increased infiltration of macrophages (Mantovani, et al. 2002; Conti and Rollins 2004; Balkwill, et al. 2005). Interestingly, CCL2 production has also been detected in TAMs, indicating the existence of an amplification loop for their recruitment (Ueno, et al. 2000; Mantovani, et al. 2002). Other CC chemokines related to CCL2, such as CCL7 and CCL8, are also produced by tumors and shown to recruit monocytes (Van Damme, et al. 1992). Along with the supposed pro-tumoral role of TAM, the local production of chemokines and the extent of TAM infiltration have been studied as prognostic factors. For example, in human breast and oesophagus cancers, CCL2 levels correlated with the extent of macrophage infiltration, lymph-node metastasis and clinical aggressiveness (Saji, et al. 2001; Azenshtein, et al. 2002). In an experimental model of non-tumorigenic melanoma, low-level of CCL2 secretion, with 'physiological' accumulation of TAM, promoted tumor formation, while high CCL2 secretion resulted in massive macrophage infiltration into the tumor mass and in its destruction (Nesbit, et al. 2001). In pancreatic cancer patients, high serum levels of CCL2 were associated with more favorable prognosis and with a lower proliferative index of tumor cells (Monti, et al. 2003). These biphasic effects of CCL2 are consistent with the 'macrophage balance' hypothesis (Mantovani, et al. 1992) and emphasize the concept that levels of macrophage infiltration similar to those observed in human malignant lesions express pro-tumor activity (Bingle, et al. 2002). A variety of other chemokines have been detected in neoplastic tissues as products of either tumor cells or stromal elements. These molecules play an important role in tumor progression by direct stimulation of neoplastic growth, promotion of inflammation and induction of angiogenesis. In spite of constitutive production of neutrophil chemotactic proteins by tumor cells, CXCL8 and related chemokines, neutrophils are not a major and obvious constituent of the leukocyte infiltrate. However, these cells, though present in minute numbers, may play a key role in triggering and sustaining the inflammatory cascade. Macrophages are also recruited by molecules other than chemokines. In particular, tumor-derived cytokines interacting with tyrosine kinase receptors, such as vascular endothelial growth factor (VEGF) and macrophage-colony stimulating factor (M-CSF) (Lin, et al. 2001; Duyndam, et al. 2002) promote macrophage recruitment, as well as macrophage survival and proliferation, the latter generally limited to murine TAM (Lin, et al. 2001; Duyndam, et al. 2002; Mantovani, et al. 2002) (Fig. 9-1). Using genetic approaches, it has been demonstrated that depletion of M-CSF markedly decreases the infiltration of macrophages at the tumor site, and this correlates with a significant delay in tumor progression. By contrast, over-expression of M-CSF by tumor cells dramatically increased macrophage recruitment and this was correlated with accelerated tumor growth (Lin, et al. 2001; Aharinejad, et al. 2002). M-CSF over-expression is common among tumors of the reproductive system, including ovarian, uterine, breast and prostate, and correlates with poor prognosis (Pollard 2004). Recently, placenta-derived growth factor (PlGF), a molecule related to VEGF in terms of structure and receptor usage, has been reported to promote the survival of TAM (Adini, et al. 2002).

9.4 TAMs Express Selected M2 Protumoral Functions

The cytokine network expressed at the tumor site plays a central role in the orientation and differentiation of recruited mononuclear phagocytes, thus contributing to direct the local immune system away from antitumor functions (Mantovani, et al. 2002). This idea is supported by both preclinical and clinical observations (Goerdt and Orfanos 1999; Bingle, et al. 2002) that clearly demonstrate an association between macrophage number/density and prognosis in a variety of murine and human malignancies.

The immuno-suppressive cytokines IL-10 and TGFβ are produced by both cancer cells (ovary) and TAMs (Mantovani, et al. 2002). IL-10 promotes the differentiation of monocytes to mature macrophages and blocks their differentiation to DC (Allavena, et al. 2000). Thus, a gradient of tumor-derived IL-10 may account for differentiation along the DC versus the macrophage pathway in different micro-anatomical localizations in a tumor. Such a situation was observed in papillary carcinoma of the thyroid, where TAMs are evenly distributed throughout the tissue, in contrast to DC which are present in the periphery (Scarpino, et al. 2000). In breast carcinoma, DC with a mature phenotype (DC-LAMP+) were localized in peri-tumoral areas, while immature DC were inside the tumor (Bell, et al. 1999). Interestingly, it was shown that Stat3 is constitutively activated in tumor cells (Wang, et al. 2004) and in diverse tumor-infiltrating immune cells (Kortylewski, et al. 2005), leading to inhibition of the production of several pro-inflammatory cytokines and chemokines and to the release of factors that suppress dendritic cell maturation. Noteworthy, ablating Stat3 in hematopoietic cells triggers an intrinsic immune-surveillance system that inhibits tumor growth and metastasis (Kortylewski, et al. 2005).

As previously discussed, IL-10 promotes the M2c alternative pathway of macrophage activation and induces TAM to express M2-related functions. Indeed, under many aspects TAM summarize a number of functions expressed by M2 macrophages, involved in tuning inflammatory responses and adaptive immunity, scavenge debris, promote angiogenesis, tissue remodelling and repair. The production of IL-10, TGFβ and PGE2 by cancer cells and TAMs (Mantovani, et al. 2002) contributes to a general suppression of antitumor activities.

TAMs are poor producers of NO (Dinapoli, et al. 1996) and, in situ in ovarian cancer, only a minority of tumors and, in these, a minority of macrophages localized at the periphery scored positive for iNOS (Klimp, et al. 2001). Moreover, in contrast to M1 polarized macrophages, TAMs have been shown to be poor producers of reactive oxygen intermediates (ROIs), consistent with the hypothesis that these cells represent a skewed M2 population (Klimp, et al. 2001).

Moreover, TAMs were reported to express low levels of inflammatory cytokines (i.e., IL-12, IL-1β, TNFα, IL-6) (Mantovani, et al. 2002). Activation of NF-κB is a necessary event promoting transcription of several pro-inflammatory genes. Our previous studies (Sica, et al. 2000) indicated that TAMs display defective NF-κB activation in response to the M1 polarizing signal LPS and we observed similar results in response to the pro-inflammatory cytokines TNFα and IL-1β (Saccani, et al. 2006) and unpublished observation). Thus, in terms of cytotoxicity and expression of inflammatory cytokines, TAMs resemble the M2 macrophages. Unexpectedly, TAMs display a high level of IRF-3/STAT-1 activation, which may be part of the molecular events promoting TAM-mediated T cell deletion (Kusmartsev and Gabrilovich 2005).

In agreement with the M2 signature, TAMs also express high levels of both the scavenger receptor-A (SR-A) (Biswas, et al. 2006) and the mannose receptor (MR) (P.Allavena, unpublished observation). Further, TAMs are poor antigen-presenting cells (Mantovani, et al. 2002).

Arginase expression in TAM has not been studied. However, it has been recently proposed that the carbohydrate-binding protein galectin-1, which is abundantly expressed by ovarian cancer (van den Brule, et al. 2003) and shows specific anti-inflammatory effects, tunes the classic pathway of L-arginine resulting in a strong inhibition of the nitric oxide production by lipopolysaccharide-activated macrophages.

Angiogenesis is an M2-associated function which represents a key event in tumor growth and progression. In several studies in human cancer, TAM accumulation has been associated with angiogenesis and with the production of angiogenic factors such as VEGF and platelet-derived endothelial cell growth factor (Mantovani, et al. 2002). More recently, in human cervical cancer, VEGF-C production by TAMs was proposed to play a role in peritumoral lymphangiogenesis and subsequent dissemination of cancer cells with formation of lymphatic metastasis (Schoppmann, et al. 2002). Additionally, TAMs participate in the pro-angiogenic process by producing the angiogenic factor thymidine phosporylase (TP), which promotes endothelial cell migration in vitro and whose levels of expression are associated with tumor neovascularization (Hotchkiss, et al. 2003). TAMs also contribute to tumor progression by producing pro-angiogenic and tumor-inducing chemokines, such as CCL2 (Vicari and Caux 2002). Moreover, TAMs accumulate in hypoxic regions of tumors and hypoxia triggers a pro-angiogenic program in these cells (see below). Therefore, macrophages recruited in situ represent an indirect pathway of amplification of angiogenesis, in concert with angiogenic molecules directly produced by tumor cells. On the anti-angiogenic side, in a murine model, GM-CSF released from a primary tumor up-regulated TAM-derived metalloelastase and angiostatin production, thus suppressing tumor growth of metastases (Dong, et al. 1998).

Finally TAMs express molecules which affect tumor cell proliferation, angiogenesis and dissolution of connective tissues. These include epidermal growth factor (EGF), members of the FGF family, TGFβ, VEGF and chemokines. In lung cancer, TAMs may favor tumor progression by contributing to stroma formation and angiogenesis through their release of PDGF, in conjunction with TGF-β1 production by cancer cells (Mantovani, et al. 2002). Macrophages can produce enzymes and inhibitors which regulate the digestion of the extracellular matrix, such as MMPs, plasmin, urokinase-type plasminogen activator (uPA) and the uPA receptor. Direct evidence has been presented that MMP-9 derived from hematopoietic cells of host origin contributes to skin carcinogenesis (Coussens, et al. 2000). Chemokines have been shown to induce gene expression of various MMPs and, in particular, MMP-9 production, along with the uPA receptor (Locati, et al. 2002). Evidence suggests that MMP-9 has complex effects beyond matrix degradation including promotion of the angiogenesis switch and release of growth factors (Coussens, et al. 2000).

The mechanisms responsible for the M2 polarization of TAM have not been completely defined yet. Recent data point to tumor (ovarian, pancreatic) -derived signals which promote M2 differentiation of mononuclear phagocytes (Hagemann, et al. 2006) and Marchesi, unpublished data).

9.5 Targeting TAM

9.5.1 *Activation*

Defective NF-κB activation in TAM correlates with impaired expression of NF-κB-dependent inflammatory functions (i.e., expression of cytotoxic mediators, NO) and cytokines, including Tumor-Necrosis Factor (TNFα, IL-1 and IL-12) (Sica, et al. 2000; Mantovani, et al. 2002; Torroella-Kouri, et al. 2005). Restoration of NF-κB activity in TAM is, therefore, a potential strategy to restore M1 inflammation and intratumoral cytotoxicity. In agreement, recent evidence indicates that restoration of an M1 phenotype in TAM may provide therapeutic benefit in tumor-bearing mice. In particular, the combination of CpG plus an anti-IL-10 receptor antibody switched infiltrating macrophages from M2 to M1 and triggered innate response debulking large tumors within 16 hours (Guiducci, et al. 2005). It is likely that this treatment may restore NF-κB activation and inflammatory functions by TAM. Moreover, TAM from STAT6$^{-/-}$ tumor-bearing mice display an M1 phenotype, with a low level of arginase and a high level of NO. As a result, these mice immunologically rejected spontaneous mammary carcinoma (Sinha, et al. 2005). These data suggest that switching the TAM phenotype from M2 to M1 during tumor progression may promote antitumor activities. In this regard, the SHIP1 phosphatase was shown to play a critical role in programing macrophage M1 versus M2 functions. Mice deficient for SHIP1 display a skewed development away from M1 macrophages (which have high inducible nitric oxide synthase levels and produce NO), towards M2 macrophages (which have high arginase levels and produce ornithine) (Rauh, et al. 2004). Finally, recent reports have identified a myeloid M2-biased cell population in lymphoid organs and peripheral tissues of tumor-bearing hosts, referred to as the myeloid suppressor cells (MSC), which are suggested to contribute to the immuno-suppressive phenotype (Bronte and Zanovello 2005). These cells are phenotipically distinct from TAM and are characterized by the expression of the Gr-1 and CD11b markers. MSC use two enzymes involved in the arginine metabolism to control T cell response: inducile nitric oxide synthase and arginase 1, which deplete the milieau of arginine, causing peroxinitrite generation, as well as lack of CD3ζ chain expression and T cell apoptosis. In prostate cancer, selective antagonists of these two enzymes were proved beneficial in restoring T cell-mediated cytotoxicity (Bronte, et al. 2005). Based on the observation that constitutively activated STAT-3 plays a pivotal role in human tumor malignancy, the discovery of selective inhibitors of this pathway presents a promising strategy with antitumor activity against human and murine cancer cells in mice (Blaskovich, et al. 2003).

9.5.2 Recruitment

Chemokines and chemokine receptors are a prime target for the development of innovative therapeutic strategies in the control of inflammatory disorders. Recent results suggest that chemokine inhibitors could affect tumor growth by reducing macrophage infiltration. Preliminary results in MCP-1/CCL2 gene targeted mice suggest that this chemokine can indeed promote progression in a Her2/neu-driven spontaneous mammary carcinoma model (Conti and Rollins 2004). Thus, available information suggests that chemokines represent a valuable therapeutic target in neoplasia. CSF-1 was identified as an important regulator of mammary tumor progression to metastasis, by regulating infiltration, survival and proliferation of TAM. Transgenic expression of CSF-1 in mammary epithelium led to the acceleration of the late stages of carcinoma and increased lung metastasis, suggesting that agents directed at CSF-1/CSF-1R activity could have important therapeutic effects (Aharinejad, et al. 2002; Pollard, 2004).

Goswami, et al. (Goswami et al. 2005) described a new role of TAM in promoting invasion of breast carcinoma cells via a CSF-1/EGF paracrine loop. Thus disruption of this circuit by blockade of either EGF receptor or CSF-1 receptor signalling may represent a new therapeutic strategy to inhibit both macrophage and tumor cell migration and invasion.

In one recent study it was found that anti-M-CSF antibodies, which interferes with TAM recruitment in different mammary carcinoma models, restored susceptibility in vivo to combination chemotherapy, implying a role of TAM in chemoresistance of tumors (Paulus, et al. 2006).

Genes of the Wnt family play a crucial role in cellular proliferation, migration and tissue patterning during embryonic development. Activation of Wnt 5a member in macrophages co-cultured with breast cancer cells, was recently shown to induce cancer cell invasion through a TNF-a-mediated induction of the MMP-7 metalloprotease. This novel circuit links the migration-regulating Wnt pathway with the proteolitic cascade, both mechanisms being indispensable for successful invasion and the Wnt antagonist dickkopf-1, inhibited cancer cell invasiveness (Pukrop, et al. 2006).

Recent results have shed new light on the links between certain TAM chemokines and genetic events that cause cancer. The CXCR4 receptor lies downstream of the vonHippel/Lindau/hypoxiainducible factor (HIF) axis. Transfer of activated ras into a cervical carcinoma line – HeLa – induces IL-8/CXCL8 production that is sufficient to promote angiogenesis and progression. Moreover, a frequent early and sufficient gene rearrangement that causes papillary thyroid carcinoma (Ret-PTC) activates an inflammatory genetic program that includes CXCR4 and inflammatory chemokines in primary human thyrocytes (Borrello, et al. 2005). The emerging direct connections between oncogenes, inflammatory mediators and the chemokine system provide a strong impetus for exploration of the anticancer potential of anti-inflammatory strategies. It was further demonstrated in non-small cell lung cancer (NSCLC) that mutation of the tumor suppressor gene PTEN results in up-regulation of HIF-1 activity and, ultimately, in HIF-1-dependent transcription of the CXCR4 gene, which provides a mechanistic basis for the up-regulation of CXCR4 expression and promotion of metastasis formation (Phillips, et al. 2005). It appears, therefore, that targeting HIF-1 activity may disrupt the HIF-1/CXCR4 pathway and affect TAM accumulation, as well as cancer cell spreading and survival. The HIF-1-inducible vascular endothelial growth factor (VEGF) is commonly produced by tumors and elicits monocyte migration. There is evidence that VEGF can significantly contribute to macrophage recruitment in tumors. Along with CSF-1, this molecule also promotes macrophage survival and proliferation. Due to the localization of TAM into the hypoxic regions of tumors, viral vectors were used to transduce macrophages with therapeutic genes, such as IFNγ, that were activated only in low oxygen conditions (Carta, et al. 2001). These works present promising approaches which use macrophages as vehicles to deliver gene therapy in regions of tumor hypoxia.

9.5.3 Angiogenesis

VEGF is a potent angiogenic factor as well as a monocyte attractant that contributes to TAM recruitment. TAMs promote angiogenesis and there is evidence that inhibition of TAM recruitment plays an important role in anti-angiogenic strategies. We found that, in addition to VEGF, the angiogenic program established by hypoxia may rely also on the increased expression of CXCR4 by TAM and endothelial cells (Schioppa, et al. 2003). Intratumoral injection of CXCR4 antagonists, such as the the bicyclam AMD3100, may potentially work as in vivo inhibitors of tumor angiogenesis. Linomide, an anti-angiogenic agent, caused significant reduction of the tumor volume, in a murine prostate cancer model, by inhibiting the stimulatory effects of TAM on tumor angiogenesis (Joseph and Isaacs 1998). Based on this, the effects of Linomide, or other anti-angiogenic drugs, on the expression of pro- and anti-angiogenic molecules by TAM may be considered valuable targets for anticancer therapy.

9.5.4 Survival

Antitumor agents with selective cytotoxic activity on monocyte-macrophages would be ideal therapeutic tools for their combined action on tumor cells and TAM. We recently reported that Yondelis (Trabectedin), a natural product derived from the marine organism *Ecteinascidia turbinata*, with potent antitumor activity (Sessa, et al. 2005) is specifically cytotoxic to macrophages and TAM, while sparing the lymphocyte subset. This compound inhibits NF-Y, a transcription factor of major importance for mononuclear phagocyte differentiation. In addition, Yondelis inhibits the production of CCL2 and IL-6 both by TAM and tumor cells (Allavena, et al. 2005). These anti-inflammatory properties of Yondelis may be an extended mechanism of its antitumor activity. Finally, pro-inflammatory cytokines (i.e., IL-1 and TNF), expressed by infiltrating leukocytes, can activate NF-κB in cancer cells and contribute to their proliferation, survival and metastasis (Balkwill and Mantovani 2001; Balkwill, et al. 2005), thus representing potential anticancer targets.

9.5.5 Matrix Remodelling

Macrophages are a major source of proteases and, interestingly, MMP-9 has been found to be preferentially expressed in M2 versus M1 cells (Martinez, et al. 2006).

TAMs produce several matrix-metalloproteases (i.e., MMP2, MMP9) which degrade proteins of the extracellular matrix, and also produce activators of MMPs, such as chemokines (de Visser, et al, 2006). Inhibition of this molecular pattern may prevent degradation of extracellular matrix, as well as tumor cell invasion and migration. TAM or neutrophil-derived proteases (i.e., MMP-9 or cathepsin B) stimulate cancer invasion and metastasis (Hanahan and Weinberg 2000; Condeelis and Pollard 2006).

The biphosphonate zoledronic acid is a prototipical MMP inhibitor. In cervical cancer this compound suppressed MMP-9 expression by infiltrating macrophages and inhibited metalloprotease activity, reducing angiogenesis and cervical carcinogenesis (Giraudo, et al. 2004). The halogenated bisphosphonate derivative chlodronate is a macrophage toxin which depletes selected macrophage populations. Given the current clinical usage of this and similar agents it is important to assess whether they have potential as TAM toxins. In support of this hypothesis, clodronate encapsulated in liposomes efficiently depleted TAM in murine teratocarcinoma and human rhabdomyosarcoma mouse tumor models, resulting in significant inhibition of tumor growth (Zeisberger, et al. 2006).

The secreted protein, acidic and rich in cysteine (SPARC), has gained much interest in cancer, being either up-regulated or down-regulated in progressing tumors. SPARC produced by macrophages present in tumor stroma can modulate collagen density, leukocyte, and blood vessel infiltration (Sangaletti, et al. 2003).

9.5.6 Effector Molecules

Cyclooxygenase (COX) is a key enzyme in the prostanoid biosynthetic pathway. COX-2 is up-regulated by activated oncogenes (i.e., ß-catenin, MET), but is also produced by TAMs in response to tumor-derived factors like mucin in the case of colon cancer. The usage of COX-2 inhibitors in the form of non-steroidal anti-inflammatory drugs is associated with reduced risk of diverse tumors (colorectal, esophagus, lung, stomach, and ovary). Selective COX-2 inhibitors are now considered as part of combination therapy (Colombo and Mantovani 2005). SOCS-1 deficiency is associated to IFN-γ-dependent spontaneous development of colorectal carcinomas. Under these conditions accumulation of aberrantly activated TAM in situ is observed and these cells account for expression of carcinogenesis related enzymes (COX2, iNOS) (Hanada, et al. 2006).

The IFN-γ-inducible enzyme indoleamine 2,3-dioxygenase is a well-known suppressor of T cell activation. It catalyzes the initial rate-limiting step in tryptophan catabolism, which leads to the biosynthesis of nicotinamide adenine dinucleotide. By depleting tryptophan from local micro-environment, indoleamine 2,3-dioxygenase (IDO) blocks activation of T lymphocytes.

Ectopic expression of IDO in tumor cells has been shown to inhibit T-cell responses (Mellor, et al. 2002). Recently it was shown that inhibition of IDO may cooperate with cytotoxic agents to elicit regression of established tumors, and may increase the efficacy of cancer immuno-therapy (Muller, et al. 2005).

Prostate cancer is strongly linked to inflammation based on epidemiological and molecular analysis (Balkwill, et al. 2005). TAMs mediate hormone resistance in prostate cancer by a nuclear receptor de-repression pathway. TAM in prostate cancer, via IL-1, convert selective androgen receptor antagonists/modulators into agonist (Zhu, et al. 2006). TNF promotes asbestos carcinogenesis by blocking death of mesothelial cells via the NF-κB pathway (Yang, et al. 2006).

Acknowledgements This work was supported by Associazione Italiana per la Ricerca sul Cancro (AIRC), Italy; European Commission; Ministero Istruzione Università e Ricerca (MIUR) and Istituto Superiore di Sanità (ISS)

References

Adini, A., Kornaga, T., Firoozbakht, F. and Benjamin, L. E. (2002). Placental growth factor is a survival factor for tumor endothelial cells and macrophages. Cancer Res 62: 2749–2752.
Aharinejad, S., Abraham, D., Paulus, P., Abri, H., Hofmann, M., Grossschmidt, K., Schafer, R., Stanley, E. R. and Hofbauer, R. (2002). Colony-stimulating factor-1 antisense treatment suppresses growth of human tumor xenografts in mice. Cancer Res 62: 5317–5724.
Allavena, P., Sica, A., Vecchi, A., Locati, M., Sozzani, S. and Mantovani, A. (2000). The chemokine receptor switch paradigm and dendritic cell migration: its significance in tumor tissues. Immunol Rev 177: 141–149.
Allavena, P., Signorelli, M., Chieppa, M., Erba, E., Bianchi, G., Marchesi, F., Olimpio, C. O., Bonardi, C., Garbi, A., Lissoni, A., de Braud, F., Jimeno, J. and D'Incalci, M. (2005).

Anti-inflammatory properties of the novel antitumor agent yondelis (trabectedin): inhibition of macrophage differentiation and cytokine production. Cancer Res 65: 2964–2971.

Azenshtein, E., Luboshits, G., Shina, S., Neumark, E., Shahbazian, D., Weil, M., Wigler, N., Keydar, I. and Ben-Baruch, A. (2002). The CC chemokine RANTES in breast carcinoma progression: regulation of expression and potential mechanisms of pro-malignant activity. Cancer Res 62: 1093–1102.

Balkwill, F., Charles, K. A. and Mantovani, A. (2005). Smoldering and polarized inflammation in the initiation and promotion of malignant disease. Cancer Cell 7: 211–217.

Balkwill, F. and Mantovani, A. (2001). Inflammation and cancer: back to Virchow? Lancet 357: 539–545.

Bell, D., Chomarat, P., Broyles, D., Netto, G., Harb, G. M., Lebecque, S., Valladeau, J., Davoust, J., Palucka, K. A. and Banchereau, J. (1999). In breast carcinoma tissue, immature dendritic cells reside within the tumor, whereas mature dendritic cells are located in peritumoral areas. J Exp Med 190: 1417–1426.

Bingle, L., Brown, N. J. and Lewis, C. E. (2002). The role of tumour-associated macrophages in tumour progression: implications for new anticancer therapies. J Pathol 196: 254–265.

Biswas, S. K., Gangi, L., Paul, S., Schioppa, T., Saccani, A., Sironi, M., Bottazzi, B., Doni, A., Bronte, V., Pasqualini, F., Vago, L., Nebuloni, M., Mantovani, A. and Sica, A. (2006). A distinct and unique transcriptional programme expressed by tumor-associated macrophages: defective NF-kB and enhanced IRF-3/STAT1 activation. Blood 107: 2112–2122.

Blaskovich, M. A., Sun, J., Cantor, A., Turkson, J., Jove, R. and Sebti, S. M. (2003). Discovery of JSI-124 (cucurbitacin I), a selective Janus kinase/signal transducer and activator of transcription 3 signaling pathway inhibitor with potent antitumor activity against human and murine cancer cells in mice. Cancer Res 63: 1270–1279.

Borrello, M. G., Alberti, L., Fischer, A., Degl'innocenti, D., Ferrario, C., Gariboldi, M., Marchesi, F., Allavena, P., Greco, A., Collini, P., Pilotti, S., Cassinelli, G., Bressan, P., Fugazzola, L., Mantovani, A. and Pierotti, M. A. (2005). Induction of a pro-inflammatory program in normal human thyrocytes by the RET/PTC1 oncogene. Proc Natl Acad Sci U S A 102: 14825–14830.

Bottazzi, B., Polentarutti, N., Acero, R., Balsari, A., Boraschi, D., Ghezzi, P., Salmona, M. and Mantovani, A. (1983). Regulation of the macrophage content of neoplasms by chemoattractants. Science 220: 210–212.

Bronte, V., Kasic, T., Gri, G., Gallana, K., Borsellino, G., Marigo, I., Battistini, L., Iafrate, M., Prayer-Galetti, T., Pagano, F. and Viola, A. (2005). Boosting antitumor responses of T lymphocytes infiltrating human prostate cancers. J Exp Med 201: 1257–1268.

Bronte, V. and Zanovello, P. (2005). Regulation of immune responses by L-arginine metabolism. Nat Rev Immunol 5: 641–654.

Carta, L., Pastorino, S., Melillo, G., Bosco, M. C., Massazza, S. and Varesio, L. (2001). Engineering of macrophages to produce IFN-gamma in response to hypoxia. J Immunol 166: 5374–5380.

Colombo, M. P. and Mantovani, A. (2005). Targeting myelomonocytic cells to revert inflammation-dependent cancer promotion. Cancer Res 65: 9113–9116.

Condeelis, J. and Pollard, J. W. (2006). Macrophages: obligate partners for tumor cell migration, invasion, and metastasis. Cell 124: 263–266.

Conti, I. and Rollins, B. J. (2004). CCL2 (monocyte chemoattractant protein-1) and cancer. Semin Cancer Biol 14: 149–154.

Coussens, L. M., Tinkle, C. L., Hanahan, D. and Werb, Z. (2000). MMP-9 supplied by bone marrow-derived cells contributes to skin carcinogenesis. Cell 103: 481–490.

Coussens, L. M. and Werb, Z. (2002). Inflammation and cancer. Nature 420: 860–867.

Desnues, B., Lepidi, H., Raoult, D. and Mege, J. L. (2005). Whipple disease: intestinal infiltrating cells exhibit a transcriptional pattern of M2/alternatively activated macrophages. J Infect Dis 192: 1642–1646.

Dinapoli, M. R., Calderon, C. L. and Lopez, D. M. (1996). The altered tumoricidal capacity of macrophages isolated from tumor-bearing mice is related to reduce expression of the inducible nitric oxide synthase gene. J Exp Med 183: 1323–1329.

Dinarello, C. A. (2005). Blocking IL-1 in systemic inflammation. J Exp Med 201: 1355–1359.

Dong, Z., Yoneda, J., Kumar, R. and Fidler, I. J. (1998). Angiostatin-mediated suppression of cancer metastases by primary neoplasms engineered to produce granulocyte/macrophage colony-stimulating factor. J Exp Med 188: 755–763.

Duyndam, M. C., Hilhorst, M. C., Schluper, H. M., Verheul, H. M., van Diest, P. J., Kraal, G., Pinedo, H. M. and Boven, E. (2002). Vascular endothelial growth factor-165 over-expression stimulates angiogenesis and induces cyst formation and macrophage infiltration in human ovarian cancer xenografts. Am J Pathol 160: 537–548.

Garlanda, C., Riva, F., Veliz, T., Polentarutti, N., Pasqualini, F., Radaelli, E., Sironi, M., Nebuloni, M., Omodeo Zorini, E., Scanziani, E. and Mantovani, A. (2007, in press). Increased susceptibility to colitis-associated cancer of mice lacking TIR8, and inhibitory members of the IL-1 receptor family. Cancer Res.

Ghassabeh, G. H., De Baetselier, P., Brys, L., Noel, W., Van Ginderachter, J. A., Meerschaut, S., Beschin, A., Brombacher, F. and Raes, G. (2006). Identification of a common gene signature for type II cytokine-associated myeloid cells elicited in vivo in different pathologic conditions. Blood 108: 575–583.

Giraudo, E., Inoue, M. and Hanahan, D. (2004). An amino-bisphosphonate targets MMP-9-expressing macrophages and angiogenesis to impair cervical carcinogenesis. J Clin Invest 114: 623–633.

Goerdt, S. and Orfanos, C. E. (1999). Other functions, other genes: alternative activation of antigen-presenting cells. Immunity 10: 137–142.

Gordon, S. (2003). Alternative activation of macrophages. Nat Rev Immunol 3: 23–35.

Goswami, S., Sahai, E., Wyckoff, J. B., Cammer, M., Cox, D., Pixley, F. J., Stanley, E. R., Segall, J. E. and Condeelis, J. S. (2005). Macrophages promote the invasion of breast carcinoma cells via a colony-stimulating factor-1/epidermal growth factor paracrine loop. Cancer Res 65: 5278–5283.

Greten, F. R., Eckmann, L., Greten, T. F., Park, J. M., Li, Z. W., Egan, L. J., Kagnoff, M. F. and Karin, M. (2004). IKKbeta links inflammation and tumorigenesis in a mouse model of colitis-associated cancer. Cell 118: 285–296.

Guiducci, C., Vicari, A. P., Sangaletti, S., Trinchieri, G. and Colombo, M. P. (2005). Redirecting in vivo elicited tumor infiltrating macrophages and dendritic cells towards tumor rejection. Cancer Res 65: 3437–3446.

Hagemann, T., Wilson, J., Burke, F., Kulbe, H., Li, N. F., Pluddemann, A., Charles, K., Gordon, S. and Balkwill, F. R. (2006). Ovarian cancer cells polarize macrophages toward a tumor-associated phenotype. J Immunol 176: 5023–5032.

Hanada, T., Kobayashi, T., Chinen, T., Saeki, K., Takaki, H., Koga, K., Minoda, Y., Sanada, T., Yoshioka, T., Mimata, H., Kato, S. and Yoshimura, A. (2006). IFNgamma-dependent, spontaneous development of colorectal carcinomas in SOCS1-deficient mice. J Exp Med 203: 1391–1397.

Hanahan, D. and Weinberg, R. A. (2000). The hallmarks of cancer. Cell 100: 57–70.

Hotchkiss, K. A., Ashton, A. W., Klein, R. S., Lenzi, M. L., Zhu, G. H. and Schwartz, E. L. (2003). Mechanisms by which tumor cells and monocytes expressing the angiogenic factor thymidine phosphorylase mediate human endothelial cell migration. Cancer Res 63: 527–533.

Joseph, I. B. and Isaacs, J. T. (1998). Macrophage role in the anti-prostate cancer response to one class of anti-angiogenic agents. J Natl Cancer Inst 90: 1648–1653.

Klimp, A. H., Hollema, H., Kempinga, C., van der Zee, A. G., de Vries, E. G. and Daemen, T. (2001). Expression of cyclooxygenase-2 and inducible nitric oxide synthase in human ovarian tumors and tumor-associated macrophages. Cancer Res 61: 7305–7309.

Koehne, C. H. and Dubois, R. N. (2004). COX-2 inhibition and colorectal cancer. Semin Oncol 31: 12–21.

Kortylewski, M., Kujawski, M., Wang, T., Wei, S., Zhang, S., Pilon-Thomas, S., Niu, G., Kay, H., Mule, J., Kerr, W. G., Jove, R., Pardoll, D. and Yu, H. (2005). Inhibiting Stat3 signalling in the hematopoietic system elicits multi-component antitumor immunity. Nat Med 11: 1314–1321.

Kusmartsev, S. and Gabrilovich, D. I. (2005). STAT1 signalling regulates tumor-associated macrophage-mediated T cell deletion. J Immunol 174: 4880–4891.
Lin, E. Y., Nguyen, A. V., Russell, R. G. and Pollard, J. W. (2001). Colony-stimulating factor 1 promotes progression of mammary tumors to malignancy. J Exp Med 193: 727–740.
Locati, M., Deuschle, U., Massardi, M. L., Martinez, F. O., Sironi, M., Sozzani, S., Bartfai, T. and Mantovani, A. (2002). Analysis of the gene expression profile activated by the CC chemokine ligand 5/RANTES and by lipopolysaccharide in human monocytes. J Immunol 168: 3557–3562.
Mantovani, A. (1999). The chemokine system: redundancy for robust outputs. Immunol Today 20: 254–257.
Mantovani, A., Bottazzi, B., Colotta, F., Sozzani, S. and Ruco, L. (1992). The origin and function of tumor-associated macrophages. Immunol Today 13: 265–270.
Mantovani, A., Sica, A., Sozzani, S., Allavena, P., Vecchi, A. and Locati, M. (2004). The chemokine system in diverse forms of macrophage activation and polarization. Trends Immunol 25: 677–686.
Mantovani, A., Sozzani, S., Locati, M., Allavena, P. and Sica, A. (2002). Macrophage polarization: tumor-associated macrophages as a paradigm for polarized M2 mononuclear phagocytes. Trends Immunol 23: 549–555.
Martinez, F. O., Gordon, S., Locati, M. and Mantovani, A. (2006). Transcriptional Profiling of the Human Monocyte-to-Macrophage Differentiation and Polarization: New Molecules and Patterns of Gene Expression. J Immunol 177: 7303–7311.
Matsushima, K., Larsen, C. G., DuBois, G. C. and Oppenheim, J. J. (1989). Purification and characterization of a novel monocyte chemotactic and activating factor produced by a human myelomonocytic cell line. J Exp Med 169: 1485–1490.
Mellor, A. L., Keskin, D. B., Johnson, T., Chandler, P. and Munn, D. H. (2002). Cells expressing indoleamine 2,3-dioxygenase inhibit T cell responses. J Immunol 168: 3771–3776.
Monti, P., Leone, B. E., Marchesi, F., Balzano, G., Zerbi, A., Scaltrini, F., Pasquali, C., Calori, G., Pessi, F., Sperti, C., Di Carlo, V., Allavena, P. and Piemonti, L. (2003). The CC chemokine MCP-1/CCL2 in pancreatic cancer progression: regulation of expression and potential mechanisms of anti-malignant activity. Cancer Res 63: 7451–7461.
Muller, A. J., DuHadaway, J. B., Donover, P. S., Sutanto-Ward, E. and Prendergast, G. C. (2005). Inhibition of indoleamine 2,3-dioxygenase, an immuno-regulatory target of the cancer suppression gene Bin1, potentiates cancer chemotherapy. Nat Med 11: 312–319.
Nesbit, M., Schaider, H., Miller, T. H. and Herlyn, M. (2001). Low-level monocyte chemoattractant protein-1 stimulation of monocytes leads to tumor formation in nontumorigenic melanoma cells. J Immunol 166: 6483–6490.
Noel, W., Raes, G., Hassanzadeh Ghassabeh, G., De Baetselier, P. and Beschin, A. (2004). Alternatively activated macrophages during parasite infections. Trends Parasitol 20: 126–133.
Paik, S., Shak, S., Tang, G., Kim, C., Baker, J., Cronin, M., Baehner, F. L., Walker, M. G., Watson, D., Park, T., Hiller, W., Fisher, E. R., Wickerham, D. L., Bryant, J. and Wolmark, N. (2004). A multi-gene assay to predict recurrence of tamoxifen-treated, node-negative breast cancer. N Engl J Med 351: 2817–2826.
Paulus, P., Stanley, E. R., Schafer, R., Abraham, D. and Aharinejad, S. (2006). Colony-stimulating factor-1 antibody reverses chemo-resistance in human MCF-7 breast cancer xenografts. Cancer Res 66: 4349–4356.
Phillips, R. J., Mestas, J., Gharaee-Kermani, M., Burdick, M. D., Sica, A., Belperio, J. A., Keane, M. P. and Strieter, R. M. (2005). Epidermal growth factor and hypoxia-induced expression of CXC chemokine receptor 4 on non-small cell lung cancer cells is regulated by the phosphatidylinositol 3-kinase/PTEN/AKT/mammalian target of rapamycin signalling pathway and activation of hypoxia inducible factor-1alpha. J Biol Chem 280: 22473–22481.
Pikarsky, E., Porat, R. M., Stein, I., Abramovitch, R., Amit, S., Kasem, S., Gutkovich-Pyest, E., Urieli-Shoval, S., Galun, E. and Ben-Neriah, Y. (2004). NF-kappaB functions as a tumour promoter in inflammation-associated cancer. Nature 431: 461–466.

Pollard, J. W. (2004). Tumour-educated macrophages promote tumour progression and metastasis. Nat Rev Cancer 4: 71–78.

Pukrop, T., Klemm, F., Hagemann, T., Gradl, D., Schulz, M., Siemes, S., Trumper, L. and Binder, C. (2006). Wnt 5a signalling is critical for macrophage-induced invasion of breast cancer cell lines. Proc Natl Acad Sci U S A 103: 5454–5459.

Rauh, M. J., Sly, L. M., Kalesnikoff, J., Hughes, M. R., Cao, L. P., Lam, V. and Krystal, G. (2004). The role of SHIP1 in macrophage programming and activation. Biochem Soc Trans 32: 785–788.

Rollins, B. J. (1997). Chemokines. Blood 90: 909–928.

Saji, H., Koike, M., Yamori, T., Saji, S., Seiki, M., Matsushima, K. and Toi, M. (2001). Significant correlation of monocyte chemoattractant protein-1 expression with neovascularization and progression of breast carcinoma. Cancer 92: 1085–1091.

Saccani, A., Schioppa, T., Porta, C., Biswas, SK., Nebuloni, M., Vago, L., Bottazzi, B., Colombo, MP., Mantovani, A., Sica, A. (2006). p50 nuclear factor-kappaB overexpression in tumor-associated macrophages inhibits M1 inflammatory responses and antitumor resistance. Cancer Res. 66(23): 11432–11440.

Sangaletti, S., Stoppacciaro, A., Guiducci, C., Torrisi, M. R. and Colombo, M. P. (2003). Leukocyte, rather than tumor-produced SPARC, determines stroma and collagen type IV deposition in mammary carcinoma. J Exp Med 198: 1475–1485.

Scarpino, S., Stoppacciaro, A., Ballerini, F., Marchesi, M., Prat, M., Stella, M. C., Sozzani, S., Allavena, P., Mantovani, A. and Ruco, L. P. (2000). Papillary carcinoma of the thyroid: hepatocyte growth factor (HGF) stimulates tumor cells to release chemokines active in recruiting dendritic cells. Am J Pathol 156: 831–837.

Schioppa, T., Uranchimeg, B., Saccani, A., Biswas, S. K., Doni, A., Rapisarda, A., Bernasconi, S., Saccani, S., Nebuloni, M., Vago, L., Mantovani, A., Melillo, G. and Sica, A. (2003). Regulation of the chemokine receptor CXCR4 by hypoxia. J Exp Med 198: 1391–1402.

Schoppmann, S. F., Birner, P., Stockl, J., Kalt, R., Ullrich, R., Caucig, C., Kriehuber, E., Nagy, K., Alitalo, K. and Kerjaschki, D. (2002). Tumor-associated macrophages express lymphatic endothelial growth factors and are related to peritumoral lymphangiogenesis. Am J Pathol 161: 947–956.

Scotton, C. J., Martinez, F. O., Smelt, M. J., Sironi, M., Locati, M., Mantovani, A. and Sozzani, S. (2005). Transcriptional profiling reveals complex regulation of the monocyte IL-1 beta system by IL-13. J Immunol 174: 834–845.

Sessa, C., De Braud, F., Perotti, A., Bauer, J., Curigliano, G., Noberasco, C., Zanaboni, F., Gianni, L., Marsoni, S., Jimeno, J., D'Incalci, M., Dall'o, E. and Colombo, N. (2005). Trabectedin for women with ovarian carcinoma after treatment with platinum and taxanes fails. J Clin Oncol 23: 1867–1874.

Sica, A., Saccani, A., Bottazzi, B., Polentarutti, N., Vecchi, A., van Damme, J. and Mantovani, A. (2000). Autocrine production of IL-10 mediates defective IL-12 production and NF-kappa B activation in tumor-associated macrophages. J Immunol 164: 762–767.

Sinha, P., Clements, V. K. and Ostrand-Rosenberg, S. (2005). Reduction of myeloid-derived suppressor cells and induction of M1 macrophages facilitate the rejection of established metastatic disease. J Immunol 174: 636–645.

Takahashi, H., Tsuda, Y., Takeuchi, D., Kobayashi, M., Herndon, D. N. and Suzuki, F. (2004). Influence of systemic inflammatory response syndrome on host resistance against bacterial infections. Crit Care Med 32: 1879–1885.

Torroella-Kouri, M., Ma, X., Perry, G., Ivanova, M., Cejas, P. J., Owen, J. L., Iragavarapu-Charyulu, V. and Lopez, D. M. (2005). Diminished expression of transcription factors nuclear factor kappaB and CCAAT/enhancer binding protein underlies a novel tumor evasion mechanism affecting macrophages of mammary tumor-bearing mice. Cancer Res 65: 10578–10584.

Tsuda, Y., Takahashi, H., Kobayashi, M., Hanafusa, T., Herndon, D. N. and Suzuki, F. (2004). Three different neutrophil subsets exhibited in mice with different susceptibilities to infection by methicillin-resistant Staphylococcus aureus. Immunity 21: 215–226.

Ueno, T., Toi, M., Saji, H., Muta, M., Bando, H., Kuroi, K., Koike, M., Inadera, H. and Matsushima, K. (2000). Significance of macrophage chemoattractant protein-1 in macrophage recruitment, angiogenesis, and survival in human breast cancer. Clin Cancer Res 6: 3282–3289.

Van Damme, J., Proost, P., Lenaerts, J. P. and Opdenakker, G. (1992). Structural and functional identification of two human, tumor-derived monocyte chemotactic proteins (MCP-2 and MCP-3) belonging to the chemokine family. J Exp Med 176: 59–65.

van den Brule, F., Califice, S., Garnier, F., Fernandez, P. L., Berchuck, A. and Castronovo, V. (2003). Galectin-1 accumulation in the ovary carcinoma peritumoral stroma is induced by ovary carcinoma cells and affects both cancer cell proliferation and adhesion to laminin-1 and fibronectin. Lab Invest 83: 377–386.

Vicari, A. P. and Caux, C. (2002). Chemokines in cancer. Cytokine Growth Factor Rev 13: 143–154.

Voronov, E., Shouval, D. S., Krelin, Y., Cagnano, E., Benharroch, D., Iwakura, Y., Dinarello, C. A. and Apte, R. N. (2003). IL-1 is required for tumor invasiveness and angiogenesis. Proc Natl Acad Sci U S A 100: 2645–2650.

Wang, T., Niu, G., Kortylewski, M., Burdelya, L., Shain, K., Zhang, S., Bhattacharya, R., Gabrilovich, D., Heller, R., Coppola, D., Dalton, W., Jove, R., Pardoll, D. and Yu, H. (2004). Regulation of the innate and adaptive immune responses by Stat-3 signalling in tumor cells. Nat Med 10: 48–54.

Wyckoff, J., Wang, W., Lin, E. Y., Wang, Y., Pixley, F., Stanley, E. R., Graf, T., Pollard, J. W., Segall, J. and Condeelis, J. (2004). A paracrine loop between tumor cells and macrophages is required for tumor cell migration in mammary tumors. Cancer Res 64: 7022–7029.

Wynn, T. A. (2004). Fibrotic disease and the T(H)1/T(H)2 paradigm. Nat Rev Immunol 4: 583–594.

Yang, H., Bocchetta, M., Kroczynska, B., Elmishad, A. G., Chen, Y., Liu, Z., Bubici, C., Mossman, B. T., Pass, H. I., Testa, J. R., Franzoso, G. and Carbone, M. (2006). TNF-alpha inhibits asbestos-induced cytotoxicity via a NF-kappaB-dependent pathway, a possible mechanism for asbestos-induced oncogenesis. Proc Natl Acad Sci U S A 103: 10397–10402.

Yang, J. and Richmond, A. (2001). Constitutive IkappaB kinase activity correlates with nuclear factor-kappaB activation in human melanoma cells. Cancer Res 61: 4901–4909.

Yoshimura, T., Matsushima, K., Oppenheim, J. J. and Leonard, E. J. (1987). Neutrophil chemotactic factor produced by lipopolysaccharide (LPS)-stimulated human blood mononuclear leukocytes: partial characterization and separation from interleukin 1 (IL-1). J Immunol 139: 788–793.

Zeisberger, S. M., Odermatt, B., Marty, C., Zehnder-Fjallman, A. H., Ballmer-Hofer, K. and Schwendener, R. A. (2006). Clodronate-liposome-mediated depletion of tumour-associated macrophages: a new and highly effective anti-angiogenic therapy approach. Br J Cancer 95: 272–281.

Zhu, P., Baek, S. H., Bourk, E. M., Ohgi, K. A., Garcia-Bassets, I., Sanjo, H., Akira, S., Kotol, P. F., Glass, C. K., Rosenfeld, M. G. and Rose, D. W. (2006). Macrophage/cancer cell interactions mediate hormone resistance by a nuclear receptor de-repression pathway. Cell 124: 615–629.

Zhu, Z., Zheng, T., Homer, R. J., Kim, Y. K., Chen, N. Y., Cohn, L., Hamid, Q. and Elias, J. A. (2004). Acidic mammalian chitinase in asthmatic Th2 inflammation and IL-13 pathway activation. Science 304: 1678–1682.

10
Clinical Development of Epidermal Growth Factor Receptor (EGFR) Tyrosine Kinase Inhibitors: What Lessons Have We Learned?

Manuel Hidalgo M.D., Ph.D.

10.1 Introduction

The epidermal growth factor receptor (EGFR) was selected as a strategic target for anticancer drug development almost two decades ago. This was based on evidence of receptor over-expression in human cancer and association with worse prognosis. Therapeutic strategies were developed and showed preclinical evidence of antitumor effects in animal models of EGFR-driven tumors. The fundamental process leading to EGFR dysregulation in human cancer were not known at that time. These agents were among the first class of targeted agents to enter the clinic at a time when the need to change the clinical development process use for cytotoxic agents to accommodate this new class of drugs was starting to be discussed. Two areas were of major interest. One was to base dose selection in pharmacodynamic endpoints rather than toxicity-based criteria. The second was to elucidate which patients are more likely to respond to these agents. Over the last few years this has been an important area of research. We have learned that while pharmacodynamic endpoints are ideal, the lack of robust and well validated analytical methods may lead to the wrong dose selection. In addition, while the average patient may benefit from these treatments, it is now clear that patients with genetic dysregulation of the EGFR by either mutations or amplifications or both are the best candidates for these treatments. It is not clear, however, how to learn about these predictors of response at earlier stages in the clinical development so that enrichment strategies can be implemented.

10.2 The HER Family of Receptors

The EGFR (HER1) is a member of the HER family of membrane receptors (HER1 through 4). The other members are HER2 (also termed ErbB2 or HER2/*neu*), HER3 (also termed ErbB3), and HER4 (also termed ErbB4). These receptors share

---

Sidney Kimmel Comprehensive Cancer Center at Johns Hopkins University School of Medicine, mhidalg1@jhmi.edu

the same molecular structure with an extracellular, cysteine-rich ligand-binding domain, a single alpha-helix transmembrane domain, and an intracellular domain with tyrosine kinase (TK) activity in the carboxy-terminal tail (excepting the HER3) (1). The TK domains of HER2 and HER4 show an 80 percent homology to that of the EGFR (2). Epidermal growth factor (EGF), transforming growth factor α (TGFα), and amphiregulin bind exclusively to the EGFR, whereas betacellulin and epiregulin bind both the EGFR and HER4. Ligand binding induces EGFR homodimerization, as well as heterodimerization with other types of HER proteins (3, 4). HER2 does not bind to any known ligand, but it is the preferred heterodimerization partner for EGFR after ligand-induced activation (5). EGFR/EGFR homodimers are unstable, whereas EGFR/HER2 heterodimers are stable, and recycle more rapidly to the cell surface (6). EGFR dimerization induces TK catalytic activity, which leads to the autophosphorylation in one or more of the five tyrosine residues in the carboxy-terminal tail, producing phosphotyrosine sites (Y992, Y1068, Y1086, Y1448, and Y1173) where adaptor and docking molecules ultimately bind (7). EGFR intracellular signalling is mainly mediated through two interrelated downstream pathways – the Ras-Raf-mitogen-activated protein kinases (MAPK, also known as extra-cytoplasmatic regulated kinases, ERK1 and ERK2), and the phosphatidylinositol 3-kinase (PI3K)/Akt pathways (8, 9). ERKs regulate the transcription of molecules involved in cell proliferation, transformation, and metastasis development (10), whereas the Akt pathway is more relevant in cell survival processes (11). An alternative route of EGFR-mediated transduction of extracellular signals is via the stress-activated protein kinase pathway that involves protein kinase C (PKC), although the basis of this regulation remains obscure. The finding that PKC has a role in EGFR transactivation and ERK regulation further complicates this regulatory mechanism (12). EGFR signaling ultimately causes increased proliferation (13), angiogenesis (14), metastasis (15), and decreased apoptosis (16). Under physiological conditions ligand binding is required to activate EGFR; however, in tumor cells there are additional mechanisms of EGFR activation. First, receptor over-expression leading to ligand-independent dimerization is commonly found in many different solid human tumors (17). Second, autocrine production of ligands (such as TGF() by tumor cells has been linked to receptor over-expression, and may represent an efficient mechanism of EGFR-driven growth (18).

10.3 Early Days Rationale to Target the EGFR

The rationale to target the EGFR for cancer therapeutics is elegantly described by Dr. John Mendelshon during his David A. Karnosky Award Lecture in 2002 (19). As described in that paper, the key concepts that led to strategies targeting the EGFR were the notion that growth factor receptors, in general, were attractive therapeutic targets, particularly for monoclonal. The EGFR was selected among many growth factor receptors because the EGFR over-expression correlates with a worse clinical outcome in several cancers including non-small-cell lung cancer (NSCLC), and tumors of the prostate, breast, stomach, colon, ovary, and head and neck, further supporting their role in tumorigenesis (17, 20, 21). It is estimated that between 40 percent and 80 percent of NSCLC over-express EGFR, and 20 to 30 percent over-express HER2 (22-24). The pivotal role that the EGFR plays as a sensor of the extracellular environment and the maintenance of cellular homeostasis makes it an *a priori* ideal candidate for a cell in transformation to exploit in order to acquire advantageous features such as freedom of movement, nutrient independence, and immortality. The EGFR was proposed as a rational target for drug development more than 20 years ago (25, 26).

10.4 Strategies to Inhibit the EGFR

Numerous classes of drugs that target the EGFR are under development, and over the last few years, an increasing number of compounds directed against the EGFR have entered clinical development and are currently in clinical trials. Two strategies have been more extensively explored in the clinic. One is when small molecules that compete with adenosine triphosphate (ATP) for binding to the receptor's kinase pocket, thus blocking receptor activation, also known as TK inhibitors (TKI). The other is when monoclonal antibodies (MAbs) are directed against the external domain of the receptor. In this article we will focus on small molecules – TKI.

TKIs compete with adenosine triphosphate (ATP) for binding to the receptor's kinase pocket, thus blocking receptor activation. A large number of TKIs are currently being evaluated. They can be classified according to their selectivity (specific agents with HER1-selective activity, as opposed to non-specific agents that target several members of the HER-family or other receptors), according to the reversibility of their interaction with their target (reversible or irreversible inhibitors) (Table 10-1). Two have received regulatory approval for use in NSCLC patients – gefitinib and erlotinib – and will be the focus of this paper.

10.5 Gefitinib

Gefitinib (Iressa, ZD1839, AstraZeneca) is an orally active, low molecular weight, synthetic quinazoline (27). Gefitinib reversibly and selectively targets the EGFR and blocks signal transduction processes implicated in the proliferation and

Table 10-1 Small molecules targeted to the EGFR. IC_{50} values represent substrate phosphorylation assays

Drug	Type	EGFR IC_{50} (μM)	HER-2 IC_{50} (μM)	Phase of development
Gefitinib (Iressa™, ZD1839)	Selective, reversible	0.02	3.7	Approved
Erlotinib (Tarceva™, OSI-774)	Selective, reversible	0.02	3.5	Approved
EKB-569	Selective, irreversible	0.04	1.2	III
Lapatinib (GW2016)	Bifunctional, reversible	0.01	0.009	Approved

Table 10-2 Results of trials of gefitinib in the first-line treatment of non-small-cell lung cancer

Chemotherapy	Biologic agent	Response rate (%)	Median survival (months)	1-year survival (%)	Ref
Gemcitabine and cisplatin	Placebo	44·8	11·1	45	(36)
	Gefitinib 250 mg	50·1	9·9	42	
	Gefitinib 500 mg	49·7	9·9	44	
Paclitaxel and carboplatin	Placebo	33·6	9·9	42	(37)
	Gefitinib 250 mg	35·0	9·8	42	
	Gefitinib 500 mg	32·1	8·7	38	

survival of cancer cells with minimal activity against other tyrosine kinases and serine/threonine kinases. Gefitinib prevents autophosphorilation of EGFR, resulting in the inhibition of downstream signalling pathways (28-30).

Phase I clinical trials of gefitinib showed a favorable toxicity profile, mostly consisting of skin toxicity and diarrhea. DLTs were observed at doses well above that at which antitumor activity was seen (31-33). Two Phase II studies have evaluated the clinical activity of gefitinib at two dose levels (250 and 500 mg) in patients with NSCLC that had failed at least one (210 patients) and at least two (216 patients) chemotherapy regimens for advanced disease, documenting response rates of 18.7 percent and 10.6 percent, respectively (34, 35). In these studies, a higher dose did not improve response rate and caused an increase in toxicity. Improvement in disease-related symptoms was significant in both trials. These results led to the regulatory approval of gefitinib (250 mg/d) as monotherapy treatment for patients with locally advanced or metastatic NSCLC refractory to platinum-based and docetaxel chemotherapy in the United States and Japan, among others. However, the addition of gefitinib to standard chemotherapy has failed to induce an improvement in response or survival in chemo-naïve NSCLC patients. Two placebo-controlled, double-blinded, Phase III randomized trials evaluating chemotherapy (either gemcitabine-cisplatin or paclitaxel-cisplatin) plus either gefitinib (250-500 mg) or placebo have rendered negative results (36, 37) (Table 10-2). A placebo-controlled Phase III study investigated the effect on survival of gefitinib as second-line or third-line treatment in 1,692 patients with locally advanced or metastatic non-small-cell lung cancer (38). The primary endpoint was survival in the overall population of patients and those with adenocarcinoma. Pre-planned subgroup analyses showed longer survival in the gefitinib group than the placebo group for never-smokers (n=375; 0.67 [0.49-0.92], p=0.012; median survival 8.9 vs 6.1 months) and patients of Asian origin (n=342; 0.66 [0.48-0.91], p=0.01; median survival 9.5 vs 5.5 months), but treatment with gefitinib was not associated with significant improvement in survival in either co-primary endpoint.

10.6 Erlotinib

Erlotinib (Tarceva, OSI-774, OSI Pharmaceuticals) is a quinazoline derivative, which reversibly inhibits the kinase activity of EGFR. It has shown in vitro and in vivo activity in preclinical trials in multiple human cancer cell lines, including ovarian, head and neck, and non-small-cell lung carcinoma (39, 40). Erlotinib has been evaluated in several Phase I studies using different doses and schedules, including weekly administration for three weeks every four weeks, and a continuous daily dosing (41, 42). The schedule that was ultimately chosen for further evaluation consists of the daily administration of 150 mg orally, with higher doses resulting in dose-limiting diarrhea and cutaneous acneiform rash (41). The cutaneous toxicity was dose-dependent, affected the face and upper trunk areas, appeared at the end of the first week of dosing and progressively recovered even in patients who continue taking the same dose of erlotinib. Other toxicities were mild to moderate and consisted of nausea and vomiting, elevation in bilirubin, headaches, and mucositis. The preliminary results of several disease-directed studies have been presented. Erlotinib has demonstrated clinical activity as a single agent in patients with NSCLC, ovarian cancer, and SCCHN (43-45). A combined analysis of the data of these Phase II studies showed that patients who developed a rash of any grade had a statistically significant longer median survival (46).

Data from two Phase III clinical trials in patients with non-small-cell lung cancer comparing standard chemotherapy regimens [cisplatin plus gemcitabine (47), and carboplatin plus paclitaxel (48)] with or without erlotinib showed that this approach failed to demonstrate a response or survival advantage. However, in a trial that randomized pretreated NSCLC patients, 2:1 to erlotinib:placebo subjects receiving the study drug survived 6.7 months compared with 4.7 months of those taking placebo ($p < 0.001$), and has been the first EGFR-targeted therapy to receive regulatory approval on the basis of prolongation of survival (49).

10.7 Insights Gained in the Role of EGFR in Cancer

In parallel to the clinical trials mentioned above and as the number of patients treated with these agents increased, a number of groups started to rationally seek factors that may be linked to the activity of the compounds. The first evidence came from clinical observations. It was known that female patients, patients of Asian origin, never -smokers and those with an adenocarcinoma type of NSCLC were the subgroups more likely to benefit from these agents. Subsequent molecular studies did not reveal the cause of this observation. This includes the link between receptor amplification and response to these agents, as well as the discovery of activating mutations of the *egfr* gene.

Abnormalities in *egfr* copy number are frequent in cancer. In a report that investigated *egfr* and EGFR expression (by fluorescent in situ hybridization [FISH] and immunohistochemistry [IHC], respectively) in 183 NSCLC patients, trisomy, polysomy and gene amplification were observed in 40 percent, 13 percent and 9 percent of the cases, respectively (50). EGFR over-expression was observed in 62 percent of the cases and correlated with increased gene copy number. Increased EGFR gene copy number detected by FISH is associated with improved survival after gefitinib therapy in patients with NSCLC (51). In this report, amplification or high polysomy of the *egfr* (documented in 33 of 102 patients) and high protein expression (observed in 58 of 98 patients) were significantly associated with better response (36% versus 3%, mean difference = 34%, 95% CI = 16.6 to 50.3; P < 0.001), disease control rate (67% versus 26%, mean difference = 40.6%, 95% CI = 21.5 to 59.7; P < 0.001), time to progression (9.0 versus 2.5 months, mean difference = 6.5 months, 95% CI = 2.8 to 10.3; P < 0.001), and survival (18.7 versus 7.0 months, mean difference = 11.7 months, 95% CI = 2.1 to 21.4; P = 0.03). Similar results regarding the correlation between *egfr* copy number and outcome were observed in a cohort of subjects with advanced bronchioalveolar carcinoma (BAC) (52). These two reports suggest FISH can be used to assess survival potential in patients treated with EGFR TKIs. In the latter subset of lung cancer subjects, no association was found between HER2 gene copy number and response or survival. Interestingly, in another report, increased HER2 copy number was also a solid marker of response to gefitinib therapy in a broader lung cancer population (53). Patients with HER2 FISH-positive tumors displayed increased expression of EGFR protein and gene gain. These findings highlight the relevance of the interplay between the HER family of receptors in the pathogenesis of cancer. In the univariate analysis of the NSCLC patients receiving erlotinib or placebo in the pivotal trial, survival was longer in the erlotinib group than in the placebo group when there was high EGFR expression (hazard ratio, 0.68; P = 0.02), or there was a high number of copies of *egfr* (hazard ratio, 0.44; P = 0.008) (54), but these correlations were not evident in the multivariate analysis.

Recent data have shown that mutations in the ATP-binding site of the *egfr* gene predict sensitivity of NSCLC patients to gefitinib (55, 56). In the report by Lynch, et al mutations were identified in the tyrosine kinase domain of the *egfr* gene in eight out of nine patients with gefitinib-responsive lung cancer, as compared with none of the seven patients with no response (P<0.001) (55). In the report by Paez, et al somatic mutations of the *egfr* gene were found in 15 of 58 unselected tumors from Japan and one of 61 from the United States (56). This phenomena was not agent- or family-specific, as it has been also documented in NSCLC patients treated with erlotinib (57), and in cell lines treated with the bifunctional (EGFR plus VEGFR2/KDR) inhibitor ZD6474 (58).

Mutations were either in-frame deletions or amino acid substitutions clustered around the ATP-binding pocket of the tyrosine kinase domain of EGFR. Remarkably, many of these deletions overlapped, sharing the deletion of four amino acids within exon 19. Other tumors had amino acid substitutions within exon 21, being particularly frequent and consistent in several reports the change from leucine to arginine at codon 858 (L858R). All mutations were heterozygous, and identical mutations were observed in multiple patients, suggesting an additive-specific gain of function. Matched normal tissue from available patients showed only the wild-type sequence, indicating that the mutations had arisen somatically during tumor formation. To further support the pathogenic role of mutations in determining the response of NSCLC to EGFR TKIs there are already reported cases where secondary mutations reverse an initial sensitivity to those agents (59).

The location of the mutations influences the sensitivity to EGFR inhibition. Gefitinib was more effective in patients with the deletion type of mutations than in patients with other mutations such as L858R (60). The response rate of patients with an exon 19 deletion and L858R were 84 percent and 71 percent, respectively, but only about half of the subjects bearing G719X had an objective response to gefitinib. In addition, patients with exon 19 deletions had a longer median survival after erlotinib or gefitinib than those with L858R (34 vs 8 months, respectively; P = 0.01) (61). In an analysis of erlotinib sensitivity using mutant constructs the order of sensitivity was exon 19 deletion = L858R > G719X > exon 20 insertion = wild-type, which is similar to the clinical observations so far (62).

In those initial, retrospective and non-consecutive analyses mutations were more prevalent in female patients with adenocarcinoma histology, and in Asian ethnic backgrounds. The report by Kosaka, et al confirmed in a systematic manner what had been described in the anecdotal initial series of NSCLC patients (63). *Egfr* mutations were not related to age or clinical stage, but there was a strong positive correlation between female gender, non-smoking status, adenocarcinoma subtype, and high degree of differentiation to mutation presence. Across all reports, independently of ethnic origin, *egfr* mutations appear almost exclusively in adenocarcinomas. It is relevant to note that as opposed to Western patterns, adenocarcinoma accounts for the majority of the NSCLC cases in Japan – as much as 70 percent in a series of resected cases (64). The actual difference in incidence of mutations between Japanese and American populations may, in part, arise from different ethiopathogenic factors mostly evidenced by profoundly dissimilar tobacco consumption, especially in women. Spontaneous mutation occurrence in predisposed histologic glandular cell subtypes, as opposed to carcinogen-induced in epithelial cells, may be behind these differential patterns.

In one of the first reports to gain further insight on the mechanistic basis of this observation, cell lines were transfected with such mutations, and mutant strains showed equivalent sensitivity to gefitinib concentrations 10-fold lower than parental cell lines (55). Differences in EGFR phosphorylation were noted and, as in transfection-induced mutated cell lines EGFR Tyr1068 phosphorylation, was more intense and also had a longer duration. These results may indicate that the mutations lower the threshold of efficacy for TKIs and thus render the EGFR susceptible to lower (clinically achievable) drug concentrations, which are suboptimal to efficaciously inhibit the receptor in the patients bearing the wild-type phenotype. As mentioned bellow, this may have explained the results of some of the clinical trials. Several reports indicate that the occurrence of EGFR mutations is an early event in carcinogenesis. Particularly, a study that analyzed mutation-positive and -negative cancers and normal adjacent mucosa showed that *egfr* mutations identical to the tumors were detected in the normal respiratory epithelium in 9 of 21 (43 percent) patients with mutant adenocarcinomas, but in none of the 16 patients without tumor mutations (65). The finding of mutations being more frequent in normal epithelium within tumor (43 percent) than in adjacent sites (24 percent) suggests a localized field effect phenomenon. In a small report in Japanese patients, *egfr* mutations were found in 12 of 19 (63 percent) of brain metastases of patients with NSCLC (66).

The same types of mutations were found in those where both primary and metastatic tissue were available, suggesting that mutation occurrence precedes systemic spread and supporting an early appearance.

A seminal report generated transgenic mice with inducible expression in type II pneumocytes of two common *egfr* mutants seen in human lung cancer (67). Both transgenic lines developed lung adenocarcinoma after sustained *egfr* mutant expression, confirming their oncogenic potential. Importantly, maintenance of these lung tumors was dependent on continued expression of the EGFR mutants and treatment with small molecule inhibitors (erlotinib or HKI-272), as well as prolonged treatment with a humanized anti-hEGFR antibody (cetuximab) which led to dramatic tumor regression. However, the pathogenic role of these mutations and its impact in downstream pathways is not completely understood. A report by Sordella, et al has shed some light in this issue, as it analyzed the differences in EGFR phosphorylation patterns in the five possible sites of the intracellular domain of the EGFR comparing mutated and wild-type NSCLC cell lines (68). Y1045 and Y1173 showed no differences, Y992 and Y1068 were more activated in mutated vs. wild-type, and Y845 was more activated in missense mutations vs. wild-type or deletion mutations. ERK status was equal in mutated vs. wild-type cell lines, probably because this signal is usually transduced via Y1173 to ras and then ERK. Phosphorylation of both Akt and STAT5 was higher in mutated vs. wild-type, as they are linked to Y992 and Y1068. These results suggest that 1) the downstream will ultimately depend on the mutation type, and 2) Akt status has questionable predictive value per se, as it fluctuates depending on the type of mutation phenotype present. It would be more informative to determine the actual subtype of EGFR phosphorylation, instead, to put into perspective the downstream scenario. In concordance to the prior data, Conde, et al determined in an analysis of the genetic and histological features of NSCLC patients that the mammalian target of rapamycin (mTOR) pathway was significantly more activated in both *egfr* and *Kras* mutants than in their wild-type counterparts (69). EGFR mutations tended to be associated with increased numbers of CA repeats and increased *egfr* gene copy numbers, but not with EGFR and caveolin-1 mRNA over-expression (70). In summary, it is increasingly evident that *egfr* mutations are oncogenic, appear early in tumorigenesis, are associated with specific signalling signatures, and induce a phenomenon of oncogene addiction that render the strains bearing them particularly sensitive to EGFR targeted therapies.

The initial reports were retrospective and, therefore, could not address the prevalence of *egfr* mutations in the general population of cancer patients. Pao and Miller reviewed this and the results are summarized in Table 10-3 (71). A consecutive series of 277 Japanese patients with NSCLC has shown a prevalence of mutations of 40 percent that were associated with female and non-smoker status, and adenocarcinoma subtype (63). A relevant aspect of this report is that *egfr* mutations were never found along with *Kras* mutations, and were more prevalent in non-smokers. In the trial that compared erlotinib with a placebo 177 samples were analyzed for *egfr* mutations, and a mutation incidence of 22 percent was documented (54). Finally, in a recently reported analysis of 860 consecutive NSCLC Italian patients

Table 10-3 Incidence of *egfr* mutations in various subgroups of NSCLC (71)

Characteristic	No. of tumors evaluated	No. of tumors with EGFR mutation	Positive for EGFR mutation (%)
Never-smokers	181	92	50.8
Smokers	434	39	9.0
Women	216	81	37.5
Men	422	55	13.0
Adenocarcinoma	453	142	31.3
Non-adenocarcinoma	306	7	2.3
East Asian	419	122	29.1
Non–East Asian	340	27	7.9
United States	262	25	9.5
Total	759	149	19.6

a global *egfr* mutation incidence of 4.5 percent was found (72). No mutations in 454 squamous carcinomas and 31 large cell carcinomas investigated were documented, and 39 were found in the series of 375 adenocarcinomas. Again e*gfr* mutations and *Kras* mutations were mutually exclusive. Bearing in mind that *Kras* serves as a downstream mediator for EGFR, the authors of the Italian report speculate that the mutually exclusive presence of *egfr* and *Kras* mutations may respond to an evolutionary paradigm where activating mutations in *egfr* are redundant if a mutation in *Kras* is already present (and vice versa). This may also help explain the striking inverse relationship of tobacco consumption and incidence of *egfr* mutations observed by this and other groups (73); it can be speculated that smoking tends to induce mutations in *Kras* that somehow prevent or make unnecessary other function-acquiring genetic changes. In addition, this downstream event seems to render EGFR-targeted therapy inefficacious, adding predictive value to its evaluation.

Few reports have addressed the independent prognostic value of *egfr* mutations. *Egfr* mutations were detected in 13 percent of 274 tumors of previously untreated patients with advanced NSCLC in the Phase III study that randomly assigned to carboplatin and paclitaxel with erlotinib or placebo. Mutation presence was associated with longer survival, irrespective of treatment ($P < .001$) (74). Whether this is directly related to the *egfr* mutation per se, or a consequence of the absence of *Kras* mutation, is unknown. Among erlotinib-treated patients, *egfr* mutations were associated with improved response rate ($P < .05$) and there was a trend toward an erlotinib benefit on time to progression ($P = .092$), but not improved survival ($P = .96$). In contrast, the Japanese report on 277 patients and the follow-up analysis of the gefitinib-treated subjects showed that whereas in patients that had not received the drug the mutational status had no significant prognostic value, the analysis of the patients that had received gefitinib revealed that the presence of the mutation had predictive value for increased survival (60, 63).

In an analysis of 90 NSCLC patients treated with gefitinib the response rate in the 17 patients harboring an *egfr* mutation was 65 percent in contrast to 13.7 percent in patients without mutation ($P < .001$) (75). Moreover, these 17 patients with EGFR mutation had significantly prolonged time to progression (21.7 v 1.8 months;

P < .001) and overall survival (30.5 v 6.6 months; P < .001) compared with the remaining 73 patients without mutation. In a recent report in 69 Korean NSCLC patients treated with gefitinib that analyzed the predictive value of several genetic and histologic parameters, there were no responders among carriers of *Kras* mutations that included two cases with concomitant *egfr* mutations (76); e*gfr* mutation presence was the only factor with predictive value in multivariate analysis. Other reports confirmed the predictive value of egfr mutations to TKIs in NSCLC patients, particularly of Asian origin (70, 77, 78). In the clinical trial that compared erlotinib with a placebo for NSCLC 325 samples were analyzed for EGFR expression and 177 samples were analyzed for *egfr* mutations (54). In contrast with other series in the multivariate analyses, adenocarcinoma (P=0.01), never having smoked (P<0.001), and expression of EGFR (P=0.03) were associated with an objective response, but survival after treatment with erlotinib was not influenced by the status of EGFR expression, the number of *egfr* copies, or *egfr* mutations (although EGFR expression and gene copy number appeared to be predictive in the univariate analysis). However, several methodological criticisms can be raised, including that mutational analysis was conducted in less than 25 percent of randomized patients, and that there is no indication that the sequencing was repeated in those positive cases (which may account for the high incidence of non-reported mutation types).

10.8 Lessons Learned

10.8.1 *Dose Selection is an Important Issue*

As mentioned above, while erlotinib has been approved for treatment of patients with chemotherapy-resistant NSCL, based on increased survival in a randomized clinical trial, gefitinib failed to do so. The reason underlying this discrepancy is not known. These two molecules are quite similar in mechanism of action and pharmacological properties. One possibility is the different population of patients and differences in clinical trial design. Another possibility is that gefitinib has been developed at a lower dose at which the concentration achieved is not sufficient to inhibit the wild-type receptor. Indeed, the Phase I clinical trials of gefitinib determined a maximum tolerated dose (MTD) of ~750 mg per day. However, lower doses of 250 and 500 mg were selected for initial exploratory trials. These doses were selected based on achievement of plasma levels sufficient to inhibit the receptor in the preclinical model as well as in pharmacodynamic effects in skin tissues (31-33). The studies, however, ignored the real unknown value of a plasma level, the fact that the skin and the tumor are not necessarily the same and that the methods used to determine the pharmacodynamic effects are not validated. A subsequent trial tested the efficacy of 250 versus 500 mg in NSCLC. The study concluded that the two doses were equivalent and selected the lower dose for definitive studies. The problem with this approach is that the IDEAL trials are indeed underpower to test an equivalenct hypothesis (34, 35). As mentioned above, susceptibility to the EGFR TKI is probable and in patients with low levels of wild-type receptor to patients with amplified mutant receptor. The efficacy of the drugs against these different situations is related to drug concentrations. Thus, it is likely that at the dose recommended for clinical trials, the intratumor levels of gefitinib are only effective to inhibit the most susceptible genotypes. Because these patients are not common, in a large clinical trial the effects may get diluted. In contrast, erlotinib was developed at the MTD of 150 mg per day. Indeed, pharmacodynamic studies with the erlotinib support the selection of that dose (79). It is possible that the higher dose results in intratumor drug levels high enough to target the wild-type receptor. While the overall benefit of the agent in wild-type patients is small, it is still better than a placebo and enough to result in a statistically different outcome. In summary, while selecting the dose for a targeted agent based on toxicity criteria is indeed unsophisticated and crude, selecting a lower dose based on non-validated endpoints may be detrimental.

10.8.2 Rushing Too Fast to Phase III Trials was not Very Productive

After the conclusion of the IDEAL trials, with a ~10 percent response rate, the INTACT trials were launched. The only rationale for these studies was the notion that EGFR inhibition exerted synergistic or additive effects with chemotherapy in preclinical models. Once the gefitinib studies were ongoing, erlotinib followed the same approach. It is likely that the decisions to initiate these trials were more commercial strategies rather than scientific rationale. The synergistic effects in preclinical models were, for the most part, based on artificially EGFR dependent tumors and interpretation is further limited by issues of dose used. In retrospect, the Phase II studies of the combinations, completed after the Phase III studies, were not impressive and – it could be argued – indicated that mayor differences was not realistic. The lesson learned is that there should be more preliminary exploratory data before large studies that consume significant resources are launched. A strategy used more frequently (everyday) is the randomizec Phase II design which provides a less biased estimation of the activity of a combination, and that may decrease the risk of negative Phase II studies.

10.8.3 Predicting Which Patients are More Likely to Respond

At the time clinical trials with EGFR TKI were launched, there was very little information as to which were the markers predictive of outcome. This information was deciphered after treatment of many patients in multiple clinical trials. The first lesson learned is that it is possible to find these predictors and that genetic-based markers are probably the more fruitful. The key question from a clinical development perspective is if trials should be conducted in selected patient populations or in general groups. One factor to consider is the rationale behind the target-drug interaction. In situations such as bcr/abl, c-kit or HER2 in which it was relatively clear that the agents would work in selected populations, the decision to base the development on such criteria was right (80, 81). For many other targeted agents, however, such knowledge is not available at the time studies commence or there are no well-validated tests to measure the target in tumor tissues. It is clear that more preclinical studies oriented to define biomarkers of response and not just activity is needed. In addition, in situations where a biomarker for patient selection is not available; every effort should be made to collect tumor tissue for translational studies. Indeed, modern clinical trials should obligatorily make an effort to collect such tissues so that when activity is observed, the cause can be explored.

References

1. Wells, A. EGF receptor. Int J Biochem Cell Biol 1999;31(6):637-43.
2. Arteaga, C. L. The epidermal growth factor receptor: from mutant oncogene in nonhuman cancers to therapeutic target in human neoplasia. J Clin Oncol 2001;19(18 Suppl):32S–40S.
3. Pinkas-Kramarski, R, Soussan, L, Waterman, H, et al. Diversification of Neu differentiation factor and epidermal growth factor signaling by combinatorial receptor interactions. Embo J 1996;15(10):2452–67.
4. Yarden, Y, Sliwkowski, M. X. Untangling the ErbB signalling network. Nat Rev Mol Cell Biol 2001;2(2):127–37.
5. Graus-Porta, D, Beerli, R. R., Daly, J. M., Hynes, N. E. ErbB-2, the preferred heterodimerization partner of all ErbB receptors, is a mediator of lateral signaling. Embo J 1997; 16(7):1647–55.
6. Worthylake, R., Opresko, L. K., Wiley, H. S. ErbB-2 amplification inhibits down-regulation and induces constitutive activation of both ErbB-2 and epidermal growth factor receptors. J Biol Chem 1999;274(13):8865–74.
7. Schlessinger, J. Cell signalling by receptor tyrosine kinases. Cell 2000;103(2):211–25.
8. Blenis, J. Signal transduction via the MAP kinases: proceed at your own RSK. Proc Natl Acad Sci U S A 1993;90(13):5889–92.
9. Burgering, B. M., Coffer, P. J. Protein kinase B (c-Akt) in phosphatidylinositol-3-OH kinase signal transduction. Nature 1995;376(6541):599–602.
10. Lewis, T. S., Shapiro, P. S., Ahn, N. G. Signal transduction through MAP kinase cascades. Adv Cancer Res 1998;74:49–139.
11. Cantley, L. C. The phosphoinositide 3-kinase pathway. Science 2002;296(5573):1655–7.
12. Tebar, F., Llado, A., Enrich, C. Role of calmodulin in the modulation of the MAPK signalling pathway and the transactivation of epidermal growth factor receptor mediated by PKC. FEBS Lett 2002;517(1-3):206–10.
13. Giordano, A., Rustum, Y. M., Wenner, C. E. Cell cycle: molecular targets for diagnosis and therapy: tumor suppressor genes and cell cycle progression in cancer. J Cell Biochem 1998;70(1):1–7.
14. de Jong, J. S., van Diest, P. J., van der Valk, P., Baak, J. P. Expression of growth factors, growth-inhibiting factors, and their receptors in invasive breast cancer. II: Correlations with proliferation and angiogenesis. J Pathol 1998;184(1):53–7.

15. Wells, A. Tumor invasion: role of growth factor-induced cell motility. Adv Cancer Res 2000;78:31–101.
16. Gibson, E. M., Henson, E. S., Haney, N., Villanueva, J., Gibson, S. B. Epidermal growth factor protects epithelial-derived cells from tumor necrosis factor-related apoptosis-inducing ligand-induced apoptosis by inhibiting cytochrome c release. Cancer Res 2002;62(2):488–96.
17. Salomon, D. S., Brandt, R., Ciardiello, F., Normanno, N. Epidermal growth factor-related peptides and their receptors in human malignancies. Crit Rev Oncol Hematol 1995;19(3):183–232.
18. Grandis, J. R., Melhem, M. F., Gooding, W. E., et al. Levels of TGF-alpha and EGFR protein in head and neck squamous cell carcinoma and patient survival. J Natl Cancer Inst 1998;90(11):824–32.
19. Mendelsohn, J. Targeting the epidermal growth factor receptor for cancer therapy. J Clin Oncol 2002;20(18 Suppl):1S–13S.
20. Woodburn, J. R. The epidermal growth factor receptor and its inhibition in cancer therapy. Pharmacol Ther 1999;82(2–3).241–50.
21. Nicholson, R. I., Gee, J. M., Harper, M. E. EGFR and cancer prognosis. Eur J Cancer 2001;37 Suppl 4:S9–15.
22. Rusch, V., Klimstra, D., Venkatraman, E., Pisters, P. W., Langenfeld, J., Dmitrovsky, E. Over-expression of the epidermal growth factor receptor and its ligand transforming growth factor alpha is frequent in resectable non-small-cell lung cancer, but does not predict tumor progression. Clin Cancer Res 1997;3(4):515–22.
23. Fontanini, G., De Laurentiis, M., Vignati, S., et al. Evaluation of epidermal growth factor-related growth factors and receptors and of neoangiogenesis in completely resected stage I-IIIA non-small-cell lung cancer: amphiregulin and microvessel count are independent prognostic indicators of survival. Clin Cancer Res 1998;4(1):241–9.
24. Shi, D., He, G., Cao, S., et al. Over-expression of the c-erbB-2/neu-encoded p185 protein in primary lung cancer. Mol Carcinog 1992;5(3):213–8.
25. Masui, H., Kawamoto, T., Sato, J. D., Wolf, B., Sato, G., Mendelsohn J. Growth inhibition of human tumor cells in athymic mice by anti-epidermal growth factor receptor monoclonal antibodies. Cancer Res 1984;44(3):1002–7.
26. Sato, J. D., Kawamoto, T., Le, A. D., Mendelsohn, J., Polikoff, J., Sato, G. H. Biological effects in vitro of monoclonal antibodies to human epidermal growth factor receptors. Mol Biol Med 1983;1(5):511–29.
27. Barker, A. J. G. K., Grundy, W., et al. Studies leading to the identification of ZD 1839 (IRESSA): an orally active, selective epidermal growth factor receptor tyrosine kinase inhibitor targeted to the treatment of cancer. Bioorg Med Chem 2001;11:1911–4.
28. Barker, A. J., Gibson, K. H., Grundy, W., et al. Studies leading to the identification of ZD1839 (IRESSA): an orally active, selective epidermal growth factor receptor tyrosine kinase inhibitor targeted to the treatment of cancer. Bioorg Med Chem Lett 2001;11(14):1911–4.
29. Anderson, N. G., Ahmad, T., Chan, K., Dobson, R., Bundred, N. J. ZD1839 (Iressa), a novel epidermal growth factor receptor (EGFR) tyrosine kinase inhibitor, potently inhibits the growth of EGFR-positive cancer cell lines with or without erbB2 overexpression. Int J Cancer 2001;94(6):774–82.
30. Ciardiello, F., Caputo, R., Bianco, R., et al. Inhibition of growth factor production and angiogenesis in human cancer cells by ZD1839 (Iressa), a selective epidermal growth factor receptor tyrosine kinase inhibitor. Clin Cancer Res 2001;7(5):1459–65.
31. Baselga, J., Rischin, D., Ranson, M., et al. Phase I safety, pharmacokinetic, and pharmacodynamic trial of ZD1839, a selective oral epidermal growth factor receptor tyrosine kinase inhibitor, in patients with five selected solid tumor types. J Clin Oncol 2002;20(21):4292–302.
32. Herbst, R. S., Maddox, A..M., Rothenberg, M. L., et al. Selective oral epidermal growth factor receptor tyrosine kinase inhibitor ZD1839 is generally well-tolerated and has activity in non-small-cell lung cancer and other solid tumors: results of a phase I trial. J Clin Oncol 2002;20(18):3815–25.

33. Ranson, M., Hammond, L. A., Ferry, D., et al. ZD1839, a selective oral epidermal growth factor receptor-tyrosine kinase inhibitor, is well tolerated and active in patients with solid, malignant tumors: results of a phase I trial. J Clin Oncol 2002;20(9):2240–50.
34. Fukuoka, M., Yano, S., Giaccone, G., et al. Multi-institutional randomized phase II trial of gefitinib for previously treated patients with advanced non-small-cell lung cancer. J Clin Oncol 2003;21(12):2237–46.
35. Kris, M. G., Natale, R. B., Herbst, R. S., et al. Efficacy of gefitinib, an inhibitor of the epidermal growth factor receptor tyrosine kinase, in symptomatic patients with non-small-cell lung cancer: a randomized trial. Jama 2003;290(16):2149–58.
36. Giaccone, G., Herbst, R. S., Manegold, C., et al. Gefitinib in combination with gemcitabine and cisplatin in advanced non-small-cell lung cancer: a phase III trial–INTACT 1. J Clin Oncol 2004;22(5):777–84.
37. Herbst, R. S., Giaccone, G., Schiller, J. H., et al. Gefitinib in combination with paclitaxel and carboplatin in advanced non-small-cell lung cancer: a phase III trial–INTACT 2. J Clin Oncol 2004;22(5):785–94.
38. Thatcher, N., Chang, A., Parikh, P., et al. Gefitinib plus best supportive care in previously treated patients with refractory advanced non-small-cell lung cancer: results from a randomized, placebo-controlled, multi-center study (Iressa Survival Evaluation in Lung Cancer). Lancet 2005;366(9496):1527–37.
39. Pollack, V. A., Savage, D. M., Baker, D. A., et al. Inhibition of epidermal growth factor receptor-associated tyrosine phosphorylation in human carcinomas with CP-358,774: dynamics of receptor inhibition in situ and antitumor effects in athymic mice. J Pharmacol Exp Ther 1999;291(2):739–48.
40. Moyer, J. D., Barbacci, E. G., Iwata, K. K., et al. Induction of apoptosis and cell cycle arrest by CP-358,774, an inhibitor of epidermal growth factor receptor tyrosine kinase. Cancer Res 1997;57(21):4838–48.
41. Hidalgo, M., Siu, L. L., Nemunaitis, J., et al. Phase I and pharmacologic study of OSI-774, an epidermal growth factor receptor tyrosine kinase inhibitor, in patients with advanced solid malignancies. J Clin Oncol 2001;19(13):3267–79.
42. Karp, D. F. D., Tensfeldt, T. G., et al. A phase I dose escalation study of epidermal growth factor receptor (EGFR) tyrosine kinase (TK) inhibitor CP-358,774 in patients (pts) with advanced solid tumors. Lung Cancer 2000;29(Suppl 1):72.
43. Perez-Soler, R., Chachoua, A., Huberman, M., et al. A phase II trial of the epidermal growth factor receptor (EGFR) tyrosine kinase inhibitor OSI-774, following platinum-based chemotherapy, in patients (pts) with advanced, EGFR-expressing, non-small-cell lung cancer (NSCLC). Proc Am Soc Clin Oncol 20: page 310, 2001 (abstr 1235). In.
44. Finkler, N., Gordon, A., Crozier, M., et al. Phase II Evaluation of OSI-774, a potent oral antagonist of the EGFR-TK in patients with advanced ovarian carcinoma. Proc Am Soc Clin Oncol 20: page 208, 2001 (abstr 831). In.
45. Soulieres, D., Senzer, N. N., Vokes, E. E., Hidalgo, M., Agarwala, S. S., Siu, L. L. Multicenter phase II study of erlotinib, an oral epidermal growth factor receptor tyrosine kinase inhibitor, in patients with recurrent or metastatic squamous cell cancer of the head and neck. J Clin Oncol 2004;22(1):77–85.
46. Clark, G., Pérez-Soler, R., Siu, L. A., Gordon, A., Santabárbara, P.. Rash severity is predictive of increased survival with erlotinib HCl. Proc Am Soc Clin Oncol page 196, 2003 (abstr 786). In.
47. Gatzemeier, U., Pluzanska, A., Szczesna, A., et al. Results of a phase III trial of erlotinib (OSI-774) combined with cisplatin and gemcitabine (GC) chemotherapy in advanced non-small- cell lung cancer (NSCLC). Proc Am Soc Clin Oncol 2004, J Clin Oncol Vol 22, No 14S (July 15 Supplement), (abstr 7010). In.
48. Herbst, R. S., Prager, D., Hermann, R., et al. TRIBUTE: a phase III trial of erlotinib hydrochloride (OSI-774) combined with carboplatin and paclitaxel chemotherapy in advanced non-small-cell lung cancer. J Clin Oncol 2005;23(25):5892–9.

49. Shepherd, F. A., Rodrigues Pereira, J., Ciuleanu, T., et al. Erlotinib in previously treated non-small-cell lung cancer. N Engl J Med 2005;353(2):123–32.
50. Hirsch, F. R., Varella-Garcia, M., Bunn, P. A., Jr., et al. Epidermal growth factor receptor in non-small-cell lung carcinomas: correlation between gene copy number and protein expression and impact on prognosis. J Clin Oncol 2003;21(20):3798–807.
51. Cappuzzo, F., Hirsch, F. R., Rossi, E., et al. Epidermal growth factor receptor gene and protein and gefitinib sensitivity in non-small-cell lung cancer. J Natl Cancer Inst 2005;97(9):643–55.
52. Hirsch, F. R., Varella-Garcia, M., McCoy, J., et al. Increased epidermal growth factor receptor gene copy number detected by fluorescence in situ hybridization associates with increased sensitivity to gefitinib in patients with bronchioloalveolar carcinoma subtypes: a Southwest Oncology Group Study. J Clin Oncol 2005;23(28):6838–45.
53. Cappuzzo, F., Varella-Garcia, M., Shigematsu, H., et al. Increased HER2 gene copy number is associated with response to gefitinib therapy in epidermal growth factor receptor-positive non-small-cell lung cancer patients. J Clin Oncol 2005;23(22):5007–18.
54. Tsao, M. S., Sakurada, A., Cutz, J. C., et al. Erlotinib in lung cancer - molecular and clinical predictors of outcome. N Engl J Med 2005;353(2):133–44.
55. Lynch, T. J., Bell, D. W., Sordella, R., et al. Activating mutations in the epidermal growth factor receptor underlying responsiveness of non-small-cell lung cancer to gefitinib. N Engl J Med 2004;350(21):2129–39.
56. Paez, J. G., Janne, P. A., Lee, J. C., et al. EGFR mutations in lung cancer: correlation with clinical response to gefitinib therapy. Science 2004;304(5676):1497–500.
57. Pao, W., Miller, V., Zakowski, M., et al. EGF receptor gene mutations are common in lung cancers from "never-smokers" and are associated with sensitivity of tumors to gefitinib and erlotinib. Proc Natl Acad Sci U S A 2004;101(36):13306–11.
58. Arao, T., Fukumoto, H., Takeda, M., Tamura, T., Saijo, N., Nishio, K. Small in-frame deletion in the epidermal growth factor receptor as a target for ZD6474. Cancer Res 2004;64(24):9101–4.
59. Kobayashi, S., Boggon, T. J., Dayaram, T., et al. EGFR mutation and resistance of non-small-cell lung cancer to gefitinib. N Engl J Med 2005;352(8):786–92.
60. Mitsudomi, T., Kosaka, T., Endoh, H., et al. Mutations of the epidermal growth factor receptor gene predict prolonged survival after gefitinib treatment in patients with non-small-cell lung cancer with postoperative recurrence. J Clin Oncol 2005;23(11):2513–20.
61. Riely, G. J., Pao, W., Pham, D., et al. Clinical course of patients with non-small-cell lung cancer and epidermal growth factor receptor exon 19 and exon 21 mutations treated with gefitinib or erlotinib. Clin Cancer Res 2006;12(3 Pt 1):839–44.
62. Greulich, H., Chen, T. H., Feng, W., et al. Oncogenic transformation by inhibitor-sensitive and -resistant EGFR mutants. PLoS Med 2005;2(11):e313.
63. Kosaka, T., Yatabe, Y., Endoh, H., Kuwano, H., Takahashi, T., Mitsudomi, T. Mutations of the epidermal growth factor receptor gene in lung cancer: biological and clinical implications. Cancer Res 2004;64(24):8919–23.
64. Watanabe, T., Hirono, T., Koike, T., et al. Registration of resected lung cancer in Niigata Prefecture. Jpn J Thorac Cardiovasc Surg 2004;52(5):225–30.
65. Tang, X., Shigematsu, H., Bekele, B. N., et al. EGFR tyrosine kinase domain mutations are detected in histologically normal respiratory epithelium in lung cancer patients. Cancer Res 2005;65(17):7568–72.
66. Matsumoto, S., Takahashi, K., Iwakawa, R., et al. Frequent EGFR mutations in brain metastases of lung adenocarcinoma. Int J Cancer 2006.
67. Ji, H., Li, D., Chen, L., et al. The impact of human EGFR kinase domain mutations on lung tumorigenesis and in vivo sensitivity to EGFR-targeted therapies. Cancer Cell 2006.
68. Sordella, R., Bell, D. W., Haber, D. A., Settleman, J. Gefitinib-sensitizing EGFR mutations in lung cancer activate anti-apoptotic pathways. Science 2004;305(5687):1163–7.
69. Conde, E., Angulo, B., Tang, M., et al. Molecular context of the EGFR mutations: evidence for the activation of mTOR/S6K signaling. Clin Cancer Res 2006;12 (3 Pt 1):710–7.

70. Taron, M., Ichinose, Y., Rosell, R., et al. Activating mutations in the tyrosine kinase domain of the epidermal growth factor receptor are associated with improved survival in gefitinib-treated chemorefractory lung adenocarcinomas. Clin Cancer Res 2005;11(16):5878–85.
71. Pao, W., Miller, V. A. Epidermal growth factor receptor mutations, small-molecule kinase inhibitors, and non-small-cell lung cancer: current knowledge and future directions. J Clin Oncol 2005;23(11):2556–68.
72. Marchetti, A., Martella, C., Felicioni, L., et al. EGFR Mutations in Non-Small-Cell Lung Cancer: Analysis of a Large Series of Cases and Development of a Rapid and Sensitive Method for Diagnostic Screening With Potential Implications on Pharmacologic Treatment. J Clin Oncol 2005;23(4):857–65.
73. Gebhardt, F., Zanker, K. S., Brandt B. Modulation of epidermal growth factor receptor gene transcription by a polymorphic dinucleotide repeat in intron 1. J Biol Chem 1999;274(19):13176–80.
74. Eberhard, D. A., Johnson, B. E., Amler, L. C., et al. Mutations in the epidermal growth factor receptor and in KRAS are predictive and prognostic indicators in patients with non-small-cell lung cancer treated with chemotherapy alone and in combination with erlotinib. J Clin Oncol 2005;23(25):5900–9.
75. Han, S. W., Kim, T. Y., Hwang, P. G., et al. Predictive and prognostic impact of epidermal growth factor receptor mutation in non-small-cell lung cancer patients treated with gefitinib. J Clin Oncol 2005;23(11):2493–501.
76. Han, S. W., Kim, T. Y., Jeon, Y. K., et al. Optimization of patient selection for gefitinib in non-small-cell lung cancer by combined analysis of epidermal growth factor receptor mutation, K-ras mutation, and Akt phosphorylation. Clin Cancer Res 2006;12(8):2538–44.
77. Shih, J. Y., Gow, C. H., Yu, C. J., et al. Epidermal growth factor receptor mutations in needle biopsy/aspiration samples predict response to gefitinib therapy and survival of patients with advanced non-small-cell lung cancer. Int J Cancer 2006;118(4):963–9.
78. Zhang, X. T., Li, L. Y., Mu, X. L., et al. The EGFR mutation and its correlation with response of gefitinib in previously treated Chinese patients with advanced non-small-cell lung cancer. Ann Oncol 2005;16(8):1334–42.
79. Malik, S. N., Siu, L. L., Rowinsky, E. K., et al. Pharmacodynamic evaluation of the epidermal growth factor receptor inhibitor OSI-774 in human epidermis of cancer patients. Clin Cancer Res 2003;9(7):2478–86.
80. Druker, B. J.. Imatinib as a paradigm of targeted therapies. Adv Cancer Res 2004;91:1–30.
81. Slamon, D. J., Leyland-Jones B, Shak S, et al. Use of chemotherapy plus a monoclonal antibody against HER2 for metastatic breast cancer that over-expresses HER2. N Engl J Med 2001;344(11):783–92.

11
GIST As the Model of Paradigm Shift Towards Targeted Therapy of Solid Tumors: Update and Perspective on Trial Design

Jaap Verweij, Caroline Seynaeve, and Stefan Sleijfer

11.1 Introduction

Gastrointestinal Stroma Tumors – or GIST – are the most common gastrointestinal (GI) sarcomas, and were only recently identified as a distinct clinical and histopathologic entity (Corless, Fletcher and Heinrich 2004). GIST have an incidence of 14.5 per million annually (comparable with chronic myeloid leukemia), and a prevalence of 129 per million (Nilsson, Bumming, Meis-Kindblom, Odén, Dortok, Bengt, Sablinska and Kindblom 2005). They constitute 0.2 percent of all GI tumors, but 80 percent of GI sarcomas. Their highest incidence is in the 40- to 60-year age group.

GIST originate from the interstitial cells of Cajal, which are fibroblast-like pacemaker cells of the gut, intercalated between intramural neurons and smooth muscle cells, and generating electrical slow waves. Recent data have shown that all GIST are malignant, but with a major variation in malignant potential (Corless, et al. 2004). Meanwhile tumor size and rate of mitotic figures have been identified as prognostic markers for malignant behavior (Nilsson, et al. 2005). The backbone of treatment for primary GIST is surgery; they are rather insensitive to radiotherapy. Once metastatic, the prognosis is poor with a median survival of only six months and total insensitivity to chemotherapy.

In 1998 GIST were found to harbor constitutively activated mutations of the membrane receptor c-KIT (Hirota, Isozaki, Moriyama, Hashimoto, Nishida, Ishiguro, Kawano, Hanada, Kurata, Takeda, Muhammad Tunio, Matsuzawa, Kanakura, Shinomura and Kitamura 1998) that is over-expressed in these tumors, and this finding has totally changed the treatment paradigm, particularly in metastatic GIST.

Erasmus University Medical Center, Daniel den Hoed Cancer Center, Department of Medical Oncology, Groene Hilledijk 301, 3075 EA ROTTERDAM, The Netherlands, j.verweij@erasmusmc.nl

11.2 The Introduction of Inhibitors of KIT

With the recognition of the important role of KIT in tumor growth of GIST, a molecular targeting became possible. KIT is a transmembrane receptor with a tyrosine kinase domain at the intracellular site of the receptor. Most relevant mutations found in GIST are located at this intracellular kinase domain (Heinrich, Corless, Demetri, Blanke, Von Mehren, Joensuu, McGreevy, Chen, Van den Abbeele, Bruker, Kiese, Eisenberg, Roberts, Singer, Fletcher, Silberman, Dimitrijevic and Fletcher 2003; Debiec-Rychter, Sciot, Le Cesne, Schlemmer, Hohenberger, Van Oosterom, Blay, Leyvraz, Stul and Casali 2006). This results in constant activation of downstream factors and finally to cell proliferation and resistance to apoptotic triggers. It so happened that the agent STI 571 (Imatinib, Glivec) was in development as an inhibitor of BCR-ABL for the treatment of chronic myeloid leukaemia (CML). Imatinib is not a selective inhibitor of BCR-ABL, but was also found to inhibit KIT and Platelet Derived Growth Factor Receptor (PDGFR). In view of this it was proposed to test Imatinib in GIST as well. Imatinib inhibits the phosphorylation at the tyrosine kinase site, and that is required for the receptor to exert its function, thereby inhibiting all downstream effects beyond phosphorylation.

After an initial successful treatment of a single patient, the first formal studies on Imatinib in GIST started in summer 2000 when the EORTC initiated a Phase I study with the agent (Van Oosterom, Judson, Verweij, Stroobants, Donato di Paola, Dimitrijevic, Martens, Webb, Sciot, Van Glabbeke, Silberman and Nielsen 2001). This study identified that the highest feasible daily dose in patients with GIST was 800 mg – twice as high as the conventional dose recommended at that time for the treatment of CML. In parallel, a consortium of investigators performed a randomized Phase II study comparing a daily dose of 400 mg to one of 600 mg (Demetri, Von Mehren, Blanke, Van den Abbeele, Eisenberg, Roberts, Heinrich, Tuveson, Singer, Janicek, Fletcher, Silverman, Silberman, Capdeville, Kiese, Peng, Simitrijevic, Druker, Corless, Fletcher and Joensuu 2002). In retrospect, given the pharmacological inter- and intra-patient variation in drug exposure, a major difference in efficacy was not to be expected between the doses. However, both the Phase I study and the Phase II studies indicated major antitumor activity for Imatinib in GIST (Van Oosterom, et al. 2001; Demetri, et al. 2002). Based on the data obtained in the randomized Phase II study Imatinib was registered for use in metastatic GIST in early 2002, at a dose of 400 mg daily. After performing its Phase I study EORTC performed a Phase II study using the daily dose of 800 mg (Verweij, Van Oosterom, Blay, Judson, Rodenhuis, Van der Graaf, Radford, Le Cesne, Hogendoorn, Di Paola, Brown and Nielsen 2003). Taken together, the various studies showed some differences in response rates (Table 11-1), and although this comparison is formally invalid, it made sense to pursue the development by performing randomized studies, comparing the standard dose of 400 mg daily, to the highest feasible dose of 800 mg daily. These two doses could also pharmacologically be considered distinctly different.

Table 11-1 Imatinib dose response in GIST in Phase I and II studies

Response	400 mg (N=73)	600 mg (N=74)	800 mg (N=40)
PR	36 (49%)	43 (58%)	27 (68%)
SD	23 (32%)	18 (24%)	8 (20%)
PD	12 (16%)	8 (11%)	5 (13%)
NE	2 (3%)	5 (7%)	0 (0%)

Table 11-2 Most frequent side effect from Imatinib per dose (Verweij 2004)

Dose	400 mg	800 mg
Anemia	76%	93%
Edema	55%	73%
Fatigue	50%	58%
Nausea	32%	45%
Anorexia	13%	26%
Vomiting	15%	21%
Rash	8%	27%

11.3 Randomized Phase III Studies of Imatinib in GIST

Two Phase III studies were scheduled. One was performed by the EORTC in collaboration with the Italian Sarcoma Group (ISG) and the Australasian Gastrointestinal Trials Group (AGITG) while the other was performed in the United States under auspices of SWOG in collaboration with ECOG and CALGB. Through intensive interaction between the steering committees of both studies, the study designs were basically made identical, the only difference being that the EORTC-ISG-AGITG study has progression-free survival as its primary endpoint, while the U.S. study was using overall survival as the primary endpoint. The EORTC-ISG-AGITG study entered 946 patients and had, meanwhile, extensively been reported (Verweij, Casali, Zalcberg, Le Cesne, Reichardt, Blay, Issels, Van Oosterom, Hogendoorn, Van Glabbeke, Bertulli and Judson 2004); to date the U.S. study has only been reported in abstract format. For this reason most of the following information is deduced from the EORTC-ISG-AGITG study.

As expected the higher dose of Imatinib yielded more toxicity (Verweij, et al. 2004). The most relevant side effects are summarized in Table 11-2. The U.S. study yielded similar results as reported in abstract format. Additionally, in the EORTC-ISG-AGITG study prognostic factors for side effects were identified based on multivariate models (Van Glabbeke, Verweij, Casali, Simes, Le Cesne, Reichardt, Issels, Judson, Van Oosterom and Blay 2006). These are listed in Table 11-3.

Based upon these prognostic factors, it was also possible to develop a model for the estimation of side effects in individual patients, to help physicians using Imatinib in GIST patients (Van Glabbeke, Verweij, Casali, Le Cesne, Hohenberger,

Table 11-3 Prognostic factors for Imatinib side effects (Van Glabbeke 2006)

Female sex
Increasing age
Worse Performance Score
Prior chemotherapy
Smaller lesions

Ray-Coquard, Schlemmer, Van Oosterom, Goldstein, Sciot, Hogendoorn, Brown, Bertulli and Judson 2005). The so-called Risk-calculator is a simple Excel file that can be downloaded from the EORTC website. (*www.eortc.be*).

The primary endpoint of the EORTC-ISG-AGITG study was progression-free survival. In the first report there appeared to be a small, albeit significant, improvement in median progression-free survival in the higher dose arm. Using historical data from the EORTC database on the use of chemotherapy in GIST, even though this was not a direct comparison, it became quite evident that Imatinib significantly improved the survival of patients as compared to survival in the pre-Imatinib era (Van Glabbeke 2005).

The preliminary data presented for the U.S. study showed almost identical results, albeit in that study the difference between 400 mg and 800 mg in terms of progression-free survival was not significant. In both studies, patients failing the 400 mg dose were allowed to cross over to the higher dose arm of the study. Obviously this may have affected the outcome between the arms in terms of overall survival. Again the results of both studies were almost identical showing no difference in terms of overall survival between the 400 mg dose and the studied higher doses (Zalcberg, Verweij, Casali, Le Cesne, Reichardt, Blay, Schlemmer, Van Glabbeke, Brown and Judson 2005).

Clearly, in a subset of patients, additional benefit can be obtained by increasing the dose from 400 to 800 mg daily at the time of first progression. So, the EORTC-ISG-AGITG study has shown that the higher dose provides a limited, yet significant, increase in progression-free survival (Verweij, et al. 2004) and crossover to 800 mg/day induces new disease stabilizations in patients progressing at 400 mg (Zalcberg, et al. 2005).

11.4 The Possible Relevance of Genotyping

Early assessments of the smaller Phase I and II studies (Heinrich, et al. 2003; Debiec-Rychter, Dumez, Judson, Wasag, Verweij, Brown, Dimitrijevic, Sciot, Stul, Vranck, Scurr, Hagemeijer, Van Glabbeke and Van Oosterom 2004) had already identified that there might be a different sensitivity of different mutations of the KIT receptor to Imatinib treatment. For this reason the EORTC-ISG-AGITG set out to retrospectively collect tumor tissue for the patients entered into the Phase III study. Paraffin tumor blocks were made available for 532 patients. Forty-seven of those were found not to have GIST at review by the expert pathology panel, and in

Table 11-4 Outcome of Imatinib dose escalation from 400 to 800 mg at first progression (N=133) (Zalcberg 2005)

Response	N	%
Partial	3	2.5
Stable Disease	36	30.3
Progression	79	66.4
Not evaluable	1	0.8
Too early	14	

Table 11-5 Frequency of KIT-mutations in GIST (Debiec-Rychter 2005)

Site of mutation	%
Exon 9	14.9
Exon 11	63.7
Exon 13	1.5
Exon 17	0.8
No KIT-mutation, but PDGF mutation	3.1
Wild-type (no mutation)	16.0

Table 11-6 Prognostic factors for response in a multivariate model (Van Glabbeke 2005)

Factor	P-value	Hazard ratio
Granulocytes	0.0014	1.052
Lesion size	0.0221	1.022
Hemoglobin	0.018	0.87
Exon 9 mutation	<0.0001	2.46
Wild-type	0.0006	1.89

108 cases the material was insufficient for further analysis. So for 377 patients, for whom treatment results were also available, genotyping could be performed. At the time of publication the median follow-up for those patients was 33 months, their three-year progression-free survival was 32 percent, three-year overall survival 58 percent, and overall response rate 55 percent (Debiec, et al. 2006). The frequency of found mutations is given in Table 11-5 with KIT-mutations occurring in exon 11 being the most common ones.

Interestingly there appeared to be a relationship between the frequency of observed specific mutations and the primary site of origin of the GIST. Exon 9 mutations were most frequently found in GIST of gastric origin, while exon 11 mutations were most frequently found in those of small bowel origin. The rapidity of onset of response was found independent of the type of mutation. However, the sensitivity to Imatinib in terms of tumor shrinkage and progression-free survival appeared to differ between the different mutation types In a multivariate analysis, patients with tumors bearing exon 11 KIT-mutations appeared to have the best outcome while exon 9 mutations emerged as an independent prognostic factor for a poor response (Table 11-6) (Debiec, et al. 2006).

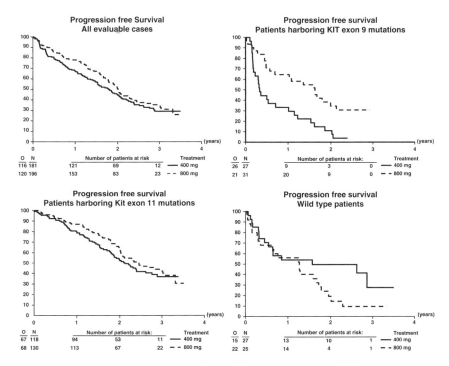

Fig. 11-1 Progression-free survival according to KIT- mutation (Debiec 2006)

Analyzing the progression-free interval that was significantly different between the two doses for the whole study, there also appeared to be a difference with respect to sensitivity to imatinib dose between tumors with different types of mutations (Fig. 11-1). While in the total group of patients with exon 11 mutations – which constitute the largest part of the GIST population – the results were similar as in the total study population; patients with exon 9 mutations appeared to do worse at the dose of 400 mg and benefited most of the dose of 800 mg. In contrast, albeit that the numbers are too small to draw firm conclusions, there is a trend towards the opposite in patients with tumors harboring wild-type KIT. Their time to progression seems less with the high dose than with the low dose. Finally, even in the subgroup of patients with exon 11 mutations, there may be a further subdivision into groups with changes in specific codons of exon 11, where tumors with changes in codons 567 and 579 seem to be less sensitive to imatinib than those with changes in other codons. If these data are confirmed in larger populations it would be the first time in history that genotyping is not only driving drug selection, but even dose selection.

Obviously, even though the results are based on a relatively large sample size, further confirmation is needed. In order to further specify these observations a meta-analysis project on the two performed Phase III studies, with the acronym Meta-GIST has recently started. The first results will become available in Summer 2007.

11.5 The Consequences for Trial Design

The above-discussed possible relationship of specific mutations having specific sensitivity for drugs or even drug doses may have very important consequences for trial design. GIST is a relatively rare disease and the subgrouping according to specific mutations will result in subgroups with a rather low incidence. If these subgroups would require specific approaches, it would become almost impossible to perform large studies. For soft tissue sarcomas, through the interaction in the Connective Tissue Oncology Society, there now is a global network emerging that might still enable scientists to perform the adequate studies, even though large Phase III studies may still be impossible. Similar problems are emerging in the field of molecularly targeted therapies in other diseases. While for Her2Neu targeting in breast cancer, a target present in 25 percent of the breast cancer population, numbers of patients in theory would still be adequate, the emerging data on the relevance of Epithelial Growth Factor Receptor (EGFR)-activating mutations in non-small-cell lung cancer are of concern with respect to trial design. The number of non-small-cell lung cancer patients with tumors harboring these mutations is small, despite the high expression of EGFR in non-small-cell lung cancer in general. If one were to pursue a study only in patients with a specific mutation of EGFR-1 in this disease, the numbers in theory may not be large enough to enable performance of a large randomized Phase III study, without appropriate global cooperation. Increasingly the targeted therapy and the need for individualized treatment will be forcing us into the direction of looking for a needle in a haystack. This is an issue we will actively have to address and solve in the years to come.

Another issue that warrants urgent attention is that the old paradigm of response may no longer be valid for drug-screening (Phase II) studies. In particular, for many of the so-called targeted agents, the preclinical models suggest tumor growth inhibition, rather than tumor regression. If a model does not show tumor regression, we should not expect such regressions in man. However, it may well be that stable disease, i.e., the absence of tumor growth for "relevant" periods of time, can also be considered as an important signal of drug activity (Van Glabbeke, Verweij, Judson and Nielsen 2002). The imatinib experience in GIST is one example. We currently know that approximately 50 percent of patients show tumor regressions, while an additional 30 to 40 percent show absence of tumor progression. The time to progression in both groups is fairly similar, so, in other words, stable disease is just as good as tumor regression in terms of progression-free survival. For soft tissue sarcomas this is not a new observation, which has been known for a long time, but unfortunately not recognized as such by the regulatory authorities. For instance the remarkable level of tumor stabilization with an agent such as Trabectidin (Yondelis, ET-743) in patients with tumors that were evidently growing prior to treatment start (Verweij and Van Glabbeke 2006), was not until now considered sufficient enough by the regulatory authority to warrant registration of the agent. Including stable disease in our assessment of drug activity will warrant the introduction of the time factor as well. Therefore, the introduction of progression-free survival at preset times is important (Van Glabbeke, 2002; LeCesne, Van Glabbeke, Verweij, Casali, Zalcberg, Reichardt Issels, Judson and Blay 2006). Yet it means that even Phase II studies may have to include a randomization, in order to exclude patient selection biases. Another way of looking at stable disease is by assessing the ratio of time to progression at the current treatment versus the preceding treatment (Mick, Crowley and Carroll 2000). Common expectation would be that the ratio is less than 1. It would be interesting if it is more than 1, and it has been proposed to consider a ratio of 1.33 or higher as evidence of drug activity in screening Phase II studies (Mick, 2000).

For imatinib in GIST, the TTP2/TTP1 ratio for the respective doses of 800 and 400 mg was more than 1.33 for at least 27 of 110 patients (Zalcberg, et al. 2005), again suggesting that the dose escalation may be beneficial to some patients.

11.6 The Importance of Continuing Drug Administration

Even though for the majority of patients' side effects of imatinib are relatively mild and infrequent, given the long duration of the induced responses and disease stabilizations it became a question of whether interrupting treatment would not be an option. The French Sarcoma Group addressed this question by performing a randomized study for patients that had not progressed after 12 months of treatment with Imatinib at a daily dose of 400 mg (LeCesne, Perol, Ray-Coquard, Bui, Duffaud, Rios, Coindre, Emile, Berthaud and Blay 2005). These patients were randomized to either continuing treatment at the same dose, or discontinuing treatment. The endpoint of the study was the time to progression after randomization. While the median progression-free survival in those patients discontinuing treatment was six months, at that point in time none of the patients that continued treatment progressed ($p = 0.0001$). Clearly, for those patients benefiting from Imatinib treatment it should be continued as long as benefit is observed.

11.7 Second Line Treatment in Metastatic Disease

Imatinib has long been considered a selective targeting agent, but given the results of the genotype relevance it is evident that the inhibition of both KIT and PDGFR play a role and, thus, the agent is a multi-targeted agent. This does not come as a surprise since cancer is a multifaceted disease and it is unlikely that a drug targeting only one factor will yield antitumor activity. The tumor itself usually has a multitude of factors at the cell membrane and in the cytoplasm that contribute to tumor growth. Furthermore, it is well known that tissue environmental factors play an important role in tumor growth as well. It is, therefore, even surprising that the relative selectivity of imatinib still yields such remarkable efficacy. But developing agents that inhibit multiple factors also made a lot of sense. More recently agents have become available that both target the tumor cell as well as angiogenetic factors. One of those agents is Sunitinib (SU011248, Sutent). It inhibits KIT and PDGFR, but also several other tyrosine kinases including VEGFR-2. After Phase I and II studies had suggested important activity for Sunitinib in GIST patients failing Imatinib or being intolerant for Imatinib, a randomized Phase III study was performed comparing Sunitinib at a dose of 50 mg daily for 28 days in a six- week cycle, to placebo, in a 2:1 randomization (Demetri, Van Oosterom, Garrett, Blackstein, Shah, Verweij, McArthur, Judson, Heinrich, Morgan, Desai, Fletcher, George, Bello, Huang, Baum, Casali 2006). The time to progression on Sunitinib was significantly longer than on placebo (28 vs. six weeks), and thus one can conclude that Sunitinib after imatinib adds an additional six months progression-free survival to the effect of imatinib. The relevance of genotype for Sunitinib treatment has yet to be studied in detail and it is conceivable that for some subsets of patients the benefits are even better (Heinrich, Maki, Corless, Antonescu, Fletcher, Fletcher, Huang, Baum and Demetri 2006).

What the data on Sunitinib also show is that there may be a whole range of other options for treatment of GIST. Many other factors have now been suggested to be involved in tumor growth and all of these are potential targets for drug treatment.

11.8 Conclusion

The introduction of the targeted agents such as Imatinib and Sunitinib has completely changed the landscape of treatment for patients with GIST. Metastatic GIST has become a treatable disease and significant life prolongation is possible. Our growing knowledge of tumor biology has enabled design of new agents with a specific focus. Interestingly, this has not only lead to better results, but even to a growing relevance of tumor genotyping since specific gene aberrations may have different sensitivity to the agents applied and even dose selection may become dependent on subtyping gene abnormalities. All of this shows that once we know the details of our target and its relationship to the disease, we can work our way to targeted treatment. In the case of imatinib, continuous exposure to the drug appears to be a crucial aspect of the activity. The effect of Sunitinib indicates that targeting multiple processes, as well as multiple tumor growth driving factors, may yield a better chance of success in malignant diseases. A problem of the developments may be that the more specific the molecular change we are aiming at is, the more difficult the trial design will become.

Based on the experiences obtained in GIST, the paradigm of treatment for this disease entity has completely shifted. Future developments, not only in GIST but also in other tumor types, will be building on this.

References

Corless, C. L., Fletcher, J. A., Heinrich, M. C. (2004) Biology of Gastrointestinal stromal tumors. J. Clin. Oncol. 22, 3813–3825.

Debiec-Rychter, M., Sciot, R., Le Cesne, A., Schlemmer, M., Hohenberger, P., Van Oosterom, A., Blay, J.-Y., Leyvraz, S., Stul, M., Casali, P. (2006) KIT-mutations and dose selection for imatinib in patients with advanced gastrointestinal stromal tumors. Eur. J. Cancer 42, 1093–1103.

Debiec-Rychter, M., Dumez, H., Judson, I., Wasag, B., Verweij, J., Brown, M., Dimitrijevic, S., Sciot, R., Stul, M., Vranck, H., Scurr, M., Hagemeijer, A., Van Glabbeke, M., and Van Oosterom, A. T. (2004) Use of c-KIT/PDGFRA mutational analysis to predict the clinical response to imatinib in patients with advanced gastrointestinal stromal tumors entered on phase I and II studies of the EORTC Soft Tissue and Bone Sarcoma Group. Eur. J. Cancer 40, 689–695.

Demetri, G. D., Von Mehren, M., Blanke, C. D., Van den Abbeele, A. D., Eisenberg, B., Roberts, P. J., Heinrich, M. C., Tuveson, D. A., Singer, S., Janicek, M., Fletcher, J. A., Silverman, S. G., Silberman, S. L., Capdeville, R., Kiese, B., Peng, B., Simitrijevic, S., Druker, B. J., Corless, C., Fletcher, C. D. M., Joensuu, H. (2002) Efficacy and safety of imatinib mesylate in advanced gastrointestinal stromal tumors. New. Engl. J. Med. 347, 472–480.

Demetri, G. D., Van Oosterom, A. T., Garrett, C. R., Blackstein, M. E., Shah, M. H., Verweij, J., McArthur, G., Judson, I. R., Heinrich, M. C., Morgan, J. A., Desai, J., Fletcher, C. D., George, S., Bello, C. L., Huang, X., Baum, C. M., Casali, P. G. (2006) Efficacy and safety of sunitinib in patients with advanced gastrointestinal stromal tumours after failure of imatinib: a randomized controlled trial. The Lancet 368, 1329–1338.

Heinrich, M. C., Corless, C. L., Demetri, G. D., Blanke, C. D., Von Mehren, M., Joensuu, H., McGreevy, L. S., Chen, C-J., Van den Abbeele, A. D., Bruker, B. J., Kiese, B., Eisenberg, B., Roberts, P. J., Singer, S., Fletcher, C. D. M., Silberman, S., Dimitrijevic, S., Fletcher, J. A. (2003) Kinase mutations and imatinib responses in patients with metastatic gastrointestinal stromal tumors. J. Clin. Oncol. 21, 4342–4349.

Heinrich, M. C., Maki R. G., Corless, C. L., Antonescu C. R., Fletcher, J. A., Fletcher, C. D., Huang, X., Baum, C. M., Demetri, G. D. (2006) Sunitinib response in imatinib-resistant GIST correlates with KIT and PDGFRA mutations. Proc. Am. Soc. Clin. Oncol. 24: abstract # 9502.

Hirota, S., Isozaki, K., Moriyama, Y., Hashimoto, K., Nishida, T., Ishiguro, S., Kawano, K., Hanada, M., Kurata, A., Takeda, M., Muhammad, Tunio G., Matsuzawa, Y., Kanakura, Y., Shinomura, Y., Kitamura, Y. (1998) Gain-of-function mutations of c-KIT in human gastrointestinal tumors. Science 279, 577–580.

Le Cesne, A., Perol, D., Ray-Coquard, I., Bui, B., Duffaud, F., Rios, M., Coindre, J. M., Emile, J. F., Berthaud, P., Blay, J.-Y. (2005) Interruption of imatinib (IM) in GIST patients with advanced disease: updated results of the prospective French Sarcoma Group randomized phase III trial on survival and quality of life. Proc. Am. Soc. Clin. Oncol. 23, abstract # 9031.

Le Cesne, A., Van Glabekke, M., Verweij, J., Casali, P., Zalcberg, J., Reichardt, P., Issels, R. D., Judson, I. R., Blay, J.-Y. (2006) Is a stable disease according to RECIST criteria a real stable disease in GIST patients treated with imatinib mesylate (IM) included in the intergroup EORTC/ISG/AGITG trial? Proc. Am. Soc. Clin. Oncol. 24, abstract # 9510.

Mick, R., Crowley, J. J., Carroll, R. J. (2000) Phase II clinical trial design for non-cytotoxic anti-cancer agents for which time to disease progression is the primary endpoint. Control. Clin. Trials. 21, 343–359.

Nilsson, B., Bumming, P., Meis-Kindblom, J. M., Odén, A., Dortok, A., Bengt, G., Sablinska, K., Kindblom, L-G. (2005) Gastrointestinal stromal tumors: the incidence, prevalence, clinical course, and prognostication in the preimatinib mesylate era. Cancer 103, 821–829.

Van Glabbeke, M., Verweij, J., Judson, I., Nielsen, O. S. (2002) Progression-free rate as the principal end point for phase II trials in soft tissue sarcomas. Eur. J. Cancer 38, 543–549.

Van Glabbeke, M., Verweij, J., Casali, P., Simes, J., Le Cesne, A., Reichardt, P., Issels, R., Judson, I., Van Oosterom, A., Blay, J.-Y. (2006) Predicting toxicities for patients with advanced gastrointestinal stromal tumors treated with imatinib: a study of the European Organisation for Research and Treatment of Cancer, the Italian Sarcoma Group, and the Australian Gastrointestinal trials group (EORTC-ISG-AGITG). Eur. J. Cancer 42, 2277–2285.

Van Glabbeke, M., Verweij, J., Casali, P. G., Le Cesne, A., Hohenberger, P., Ray-Coquard, I., Schlemmer, M., van Oosterom, A. T., Goldstein, D., Sciot, R., Hogendoorn, P., Brown, M., Bertulli, R., Judson, I. R. (2005) Initial and late resistance to Imatinib (IM) in advanced gastrointestinal stromal tumors (GIST) are predicted by different prognostic factors. An EORTC-AGITG study. J. Clin. Oncol. 23, 5795–5804.

Van Oosterom, A. T., Judson, I., Verweij J., Stroobants, S., Donato di Paola, E., Dimitrijevic, S., Martens, M., Webb, A., Sciot, R., Van Glabbeke, M., Silberman, S., Nielsen, O. S. (2001) Safety and efficacy of imatinib (STI571) in metastatic gastrointestinal stromal tumors: a phase I study. Lancet 358, 1421–1423.

Verweij, J., Van Oosterom, A., Blay, J.-Y., Judson, I., Rodenhuis, S., Van der Graaf, W., Radford, J., Le Cesne, A., Hogendoorn, P. C. W., Di Paola, E. D., Brown, M. and Nielsen, O. S. (2003) Imatinib mesylate (STI-571 Glivec, Gleevec) is an active agent for gastrointestinal stromal tumors, but does not yield responses in other soft tissue sarcomas that are unselected for a molecular target. Resultsfrom an EORTC Soft Tissue and Bone Sarcoma phase II study. Eur. J. Cancer 39, 2006–2011.

Verweij, J. and Van Glabbeke, M. (2006) Translating targets into treatment: Changes in trial methodology and treatment approaches for soft tissue sarcomas. ASCO Educational Book pp 522–530.

Verweij, J., Casali, P. G., Zalcberg, J., Le Cesne, A., Reichardt, P., Blay, J.-Y., Issels, R., Van Oosterom, A., Hogendoorn, P., Van Glabbeke, M., Bertulli, R. and Judson, I. (2004) Progression-free survival in gastrointestinal stromal tumors with high-dose imatinib: a randomized trial. Lancet 364, 1127–1134.

Zalcberg, J. R., Verweij, J., Casali, P. G., Le Cesne, A., Reichardt, P., Blay, J.-Y., Schlemmer, M., Van Glabbeke, M., Brown, M. and Judson, I. R. (2005) Outcome of patients with advanced gastrointestinal stromal tumors crossing over to a daily imatinib dose of 800 mg after progression on 400 mg. Eur. J. Cancer 41, 1751–1757.

12
Monoclonal Antibodies in the Treatment of Malignant Lymphomas

Bertrand Coiffier

12.1 Introduction

Non-Hodgkin's lymphoma is a heterogeneous group of B- and T-cell cancers, with a large variety of patterns of growth, clinical presentations, and responses to treatment. Outcome depends on histological subtype, tumor characteristics, host responses, and treatment. About 90 percent of lymphomas have a B-cell phenotype and for them recent therapeutic progress came from the introduction of monoclonal antibodies (MAb) alone or in combination with chemotherapy (Coiffier 2005; Coiffier 2005; Traulle and Coiffier 2005). The first antigen that has been targeted for therapeutic purpose with success was the CD20 antigen, a transmembrane protein expressed by more than 99 percent of B-cell lymphomas. Rituximab was the first MAb engineered to target the CD20 antigen and first approved MAb for the treatment of lymphoma patients. Through the last 10 years clinical trials with rituximab have confirmed its efficacy in follicular lymphoma, as well as in aggressive lymphomas, and its use has expanded significantly beyond the initial indication of indolent B-cell lymphomas to virtually any CD20-positive lymphoma. In recent years, several other MAb targeting CD20 or other lymphocyte antigens appeared, some of them associated with a toxin or a radioisotope.

12.2 Mechanisms of Action of Monoclonal Antibodies

The mechanisms of action of MAb differ with the type of antibody, the antigen they target, and their use: alone, in combination with chemotherapy, conjugated to a toxin or a radionucleide. In case of a naked antibody different mechanisms have been identified (Cartron, Watier, Golay, and Solal-Celigny 2004). CD20 binding by rituximab is followed by homotypic aggregation, rapid translocation of CD20 into

---

Hematology Department, Hospices Civils de Lyon and Claude Bernard University,
Address: Professeur B Coiffier, Hematology, CH Lyon-Sud, 69495 Pierre-Benite, France,
Tel +33 478 984300, Fax +33 478 864355, bertrand.coiffier@chu-lyon.fr

F. Colotta and A. Mantovani (eds.), *Targeted Therapies in Cancer*.
© Springer 2008

specialized plasma membrane microdomains known as rafts, and induction of apoptosis. Membrane rafts concentrate src family kinases and other signalling molecules (phospholipases, caspases …), and the anti-CD20-induced apoptotic signals are thought to occur as a consequence of CD20 accumulation in rafts (Janas, Priest, Wilde, White and Malhotra 2005). The role of complement-dependent cytotoxicity (CDC) is suggested by the consumption of complement observed after rituximab administration, but in vitro CDC does not correlate with clinical response in lymphomas (Weng and Levy, 2001; Winkler, Jensen, Manzke, Schulz, Diehl and Engert 1999). However, CDC seems to be the most important mechanism of cell lysis in chronic lymphocytic leukemia (CLL) patients (Kennedy, Beum, Solga, DiLillo, Lindorfer, Hess, Densmore, Williams and Taylor 2004). CDC is probably involved in the cytokine-release syndrome and its toxicity (Bienvenu, Chvetzoff, Salles, Balter, Tilly, Herbrecht, Morel, Lederlin, Solal-Celigny, Audhuy, Christian, Gabarre, Casasnovas, Marit, Sebban and Coiffier 2001). The importance of antibody-dependent cellular cytotoxicity (ADCC) has been demonstrated in vivo when rituximab is used alone (Cartron, Dacheux, Salles, Solal-Celigny, Bardos, Colombat and Watier 2002). The Fc receptor (FcγR) of effector cells has two alleles and the valine/valine (V/V) allele of FcγRIIIa which confers a higher affinity for IgG1 and rituximab is associated with an increased responsiveness to rituximab (Cartron 2002; Weng and Levy 2003). If the clinical relevance of the FcγRIIIa receptor dimorphism was established in a number of studies with rituximab used alone, it does not seem to play a major role when rituximab is used in combination with chemotherapy (Boettcher, Pott, Ritgen, Hiddemann, Unterhalt and Kneba 2004) even if one study showed an increased response for patients with the V/V allele without difference for progression-free survival (PFS) or overall survival (OS) (Kim, Jung, Kim, Lee, Yang, Park, Do, Shin, Kim, Hyun and Sohn 2006).

Finally, evidences that rituximab could synergize with chemotherapeutic agents in B-cell killing were provided by Demidem (Demidem, Lam, Alas, Hariharan, Hanna and Bonavida 1997). Subsequent investigations have confirmed synergy of rituximab with fludarabine, doxorubicin and other anticancer drugs (Alas and Bonavida 2001; Alas, Bonavida and Emmanouilides 2000; Ghetie, Bright and Vitetta 2001). In one hypothesis, this synergism is mediated, at least in part, via down-regulation of interleukin-10 (IL-10) by rituximab, which in turn causes down-regulation of the anti-apoptotic protein bcl2 and increased sensitivity to apoptosis (Vega, Huerta-Yepaz, Garban, Jazirehi, Emmanouilides and Bonavida 2004). Another mechanism involves the inhibition of the activity of P-glycoprotein and, thus, the efflux of drugs like doxorubicin or vincristine (Ghetie, Crank, Kufert, Pop and Vitetta 2006). In cell lines, the P-glycoprotein pump is translocated out of the lipids rafts. This activity seems independent of the classical antiproliferative effect of rituximab (Ghetie 2006).

If these mechanisms may have a role when MAb are combined with a radionuclide, most of the antitumor effect resides in their capacity to deliver local radiotherapy after the MAb is attached to tumor cells (Lemieux and Coiffier 2005). The choice of the antibody and therapeutic radioisotopes are critical for the success of radioimmunotherapy (RIT). Several radiolabeled MAb have been studied in

Table 12-1 Characteristics of the two registered radiolabeled MAb. Adapted from references (Cheson 2003; Dillman 2002)

	^{90}Y-Ibritumomab	^{131}I-Tositumomab
Linker	Tiuxetan	None
Isotope radiation decay	Beta	Beta and gamma
Half-life, days	2.7	8.0
Path length, mm	5.0	0.8
Energy, MeV	2.3	0.61
Non-tumor distribution	Bone	Thyroid
Urine excretion	Limited	Substantial
Imaging	Not possible	Possible

clinical trial, but only two – yttrium-90 (^{90}Y or Y-90) ibritumomab tiuxetan and iodine-131 (^{131}I or I-131) tositumomab – have been registered for the treatment of lymphoma patients. Both radiolabeled antibodies are mouse antibodies reacting with CD20-expressing tumors. Yttrium-90 is a pure β-emitter, with a half-life of 2.7 days (Cheson 2003). It is linked to the antibody by a chelator (tiuxetan). The long pathlength of its β-particles is particularly advantageous in tumors with heterogeneous or low distribution of the antigen (Juweid 2002). Iodine-131 is an α and β-emitter and has a half-life of 8.0 days. The pathlenght of its β-particles is relatively shorter then Y-90. Table 12-1 presents the differences between Y-90 and I-131 radiolabeled antibodies.

An alternative approach to increase the activity of MAbs has been the development of immunotoxin, a construct conjugating the antibody to cytotoxic plant, bacterial toxic proteins, or chemotherapy drugs (doxorubicin) (Rosenblum 2004). The commonly used toxins, ricin or diphtheria toxin, are highly potent natural products that disrupt protein synthesis. Unlike unconjugated MAb, immunotoxins must be internalized after antigen binding to allow the toxin access to the cytosol. Although the conjugation to MAbs confers some target specificity, the toxin continues to mediate non-specific toxicity to normal tissues. Deglycosylated ricin A-chain has been used to eliminate such non-specific toxicity.

12.3 Mechanisms of Resistance

If multiple mechanisms of rituximab action have been reported, it remains unclear which is/are most important in patients and, therefore, it is difficult to know the relative importance of potential mechanisms of resistance. This is true for the other MAb too. Conceptual approaches of resistance mechanisms may be resumed as followed (Smith 2003).

Concerning events up to antigen binding, resistance to rituximab may be secondary to low serum levels or rapid metabolism of the MAb; development of anti-monoclonal antibodies (HAMA), most frequent with non-humanized antibodies than with rituximab, or anti-chimeric (HACA) antibodies (not yet demonstrated in patients);

possibly different distribution within malignant nodes, blood cells, marrow and extranodal sites and responsible for poor tumor penetration; high level of soluble antigen target (not yet demonstrated for CD20 antigen); high tumor burden, and poor surface antigen expression.

Events that may induce resistance to rituximab after the antigen binding are alteration of induced intracellular signals; reduction of direct apoptosis effect in cases of elevated bcl-2 protein; inhibition of CDC by complement inhibitors, and alteration of cell-mediated immunity. Gene microarray analysis has shown that patients who failed to respond to rituximab have altered patterns of gene expression, with an over-expression of genes important in cell-mediated immunity (Bohen, Troyanskaya, Alter, Warnke, Botstein, Brown and Levy 2003).

12.4 Safety and Tolerability

The safety of MAb is mainly related to its origin and to the compound attached to it. Radiolabeled MAb has a greater hematological toxicity than naked MAb because of the effect of surrounding normal hematopoietic cells in bone marrow. Immunotoxins also have a greater toxicity because of the release of the toxin. Some MAb, such as alemtuzumab, may have a larger hematological toxicity because the target antigen (CD52 in case of alemtuzumab) is not restricted to lymphoid cells.

The safety of rituximab is mainly related to infusion toxicity, a toxicity most MAb have in common (Kimby 2005). These side effects are observed during the infusion or in the first hours after drug infusion and particularly for the first infusion. They include fever, chills, dizziness, nausea, pruritus, throat swelling, cough, fatigue, hypotension and transient bronchospasm in a majority of patients. These symptoms are part of the cytokine-release syndrome. Their intensity correlates with the number of circulating malignant cells at time of infusion. More severe infusional toxicity includes bronchospasm, angioedema and acute lung injury, often associated with high circulating cell counts or pre-existing cardiac or pulmonary disease.

Another common toxicity is the rapid depletion of normal antigen-positive B-lymphocytes from blood, bone marrow and lymph nodes of the recipient, lasting between six and nine months following the last administration of rituximab. In the case of short rituximab treatment, this depletion does not compromise immunity; immunoglobulins do not decrease significantly, and patients do not have an increased risk for infections during and after rituximab therapy (Grillo-Lopez, Hedrick, Rashford and Benyunes 2002; Kimby 2005); except for some viruses like herpes virus, cytomegalovirus, or hepatitis B virus. Maintenance treatment, particularly after autologous transplant, might be associated with a decrease in immunoglobulins (Lim, Zhang, Wang, Esler, Beggs, Pruitt, Hancock and Townsend 2004) and late toxicity (Kimby 2005).

Rare toxic events associated with rituximab comprised delayed neutropenia and pulmonary reactions. Delayed neutropenia usually occurs in patients treated with rituximab alone or in combination with chemotherapy. It appears between one and six months after the last infusion, may be transient, rarely associated with infection, and resolve spontaneously in most of the cases (Lemieux, Tartas, Traulle, Espinouse, Thieblemont, Bouafia, Alhusein, Antal, Salles and Coiffier 2004). The mechanisms are not fully understood. Pulmonary reactions are rare and diverse, usually related to rituximab because of the temporal relation (Kimby 2005) Radioimmunotherapy is associated with secondary myelodysplastic syndromes.

12.5 Clinical Studies

A few MAb have been registered for the treatment of lymphoma patients: rituximab (Rituxan® or MabThera®), ^{90}Y-ibritumomab tiuxetan (Zevalin®), ^{131}I-tositumomab (Bexxar®), and denileukin diftitox (OnTak®), the last two only in the United States. However, a lot of other MAb are currently in preclinical studies (Phase I or Phase II). Rituximab is certainly the MAb where the largest experience exists and the MAb with several demonstrative randomized studies. We will focus on demonstrated activity (Phase III studies) and some Phase II studies with promising results.

12.6 Rituximab in follicular lymphoma

Rituximab alone in relapse. When used alone, rituximab is usually given as four weekly injections of 375 mg/m^2 (Maloney, 1999). The pivotal multi-center Phase II study that included 166 patients treated with four infusions of rituximab showed an overall remission rate of 48 percent (including 6% of CR), and a median time to progression of 13 months (McLaughlin, Grillolopez, Link, Levy, Czuczman, Williams, Heyman, Bencebruckler, White, Cabanillas, Jain, Ho, Lister, Wey, Shen and Dallaire 1998). Elevated ß2-microglobulin, elevated LDH, bulky disease, and age >60 years did not appear to impact response, implying that patients regarded as having a poor prognosis may respond to rituximab.

Patients relapsing after initial response to rituximab treatment may be re-treated with comparable response rates and adverse side effects, but, interestingly, median time for progression might be longer than after first treatment (Davis, Grillo-Lopez, White, McLaughlin, Czuczman, Link, Maloney, Weaver, Rosenberg and Levy 2000; Lemieux, Bouafia, Thieblemont, Hequet, Arnaud, Tartas, Traulle, Salles and Coiffier 2004).

Whether prolonged treatment with rituximab or maintenance is able to further improve response rates and to prolong remission duration is of considerable interest. Several arguments are in favor of this approach: the success of re-treatment or the strong correlation between rituximab plasma levels and response rates (Berinstein, Grillolopez, White, Bencebruckler, Maloney, Czuczman, Green, Rosenberg, McLaughlin and Shen 1998). A recent randomized trial showed that adding maintenance doses of rituximab prolonged response duration (Ghielmini, Schmitz, Cogliatti, Pichert, Hummerjohann, Waltzer, Fey, Betticher, Martinelli, Peccatori, Hess, Zucca, Stupp, Kovacsovics, Helg, Lohri, Bargetzi, Vorobiof and Cerny 2004). Two-hundred-two patients with newly diagnosed or refractory/relapsed follicular lymphoma were treated with rituximab. Patients responding, or with stable disease, were randomized to no further treatment or prolonged rituximab administration (375 mg/m^2 every two months for four times). With a median follow-up of 35 months, the median event-free survival (EFS) was prolonged in the treated group, 23 months vs. 12 months in the control group. However, patients relapsed within the six months after stopping rituximab treatment. In another randomized study, Hainsworth showed that re-treatment at relapse or prolonged treatment have the same benefit in terms of duration of rituximab efficacy or time to chemotherapy (Hainsworth, Litchy, Shatter, Lackey, Grimaldi and Greco 2005).

Several questions remain without clear response: what is the optimal prolonged treatment? What is the optimal duration maintenance? Which patients benefit from prolonged treatment? And, finally, is prolonged treatment or re-treatment at progression better in terms of survival or impact on transformation rate?

Rituximab alone in Untreated Follicular Lymphoma. Usually, patients with no adverse prognostic factors are not treated until they develop such adverse parameters (Ardeshna, Smith, Norton, Hancock, Hoskin, MacLennan, Marcus, Jelliffe, Hudson, and Linch 2003) However, because of its low profile toxicity, its presumed low rate of secondary malignancy, and its lack of stem cell toxicity, rituximab single-agent was investigated in this setting: (Colombat, Salles, Brousse, Eftekhari, Soubeyran, Delwail, Deconinck, Haioun, Foussard, Sebban, Stamatoullas, Milpied, Boue, Taillan, Lederlin, Najman, Thieblemont, Montestruc, Mathieu-Boue, Benzohra and Solal-Celigny 2001) in a series of 50 patients, an RR of 73 percent was obtained, with 26 percent of CR; 57 percent of the informative patients in CR reached a molecular remission. However, even patients in CR and in molecular response did not seem to benefit from this treatment because the median time to progression was only two years, not longer than without treatment. A randomized study is currently underway in the United Kingdom challenging this finding in these otherwise "watch and wait" patients.

Rituximab alone was also studied in patients with a more aggressive presentation, needing treatment at diagnosis, or after some follow-up without treatment (Hainsworth, Litchy, Burris, Scullin, Corso, Yardley, Morrissey and Greco 2002). The RR, just after four infusions, was comparable with the one observed in relapsing patients (50% and less than 10% of CR). About 10 percent of patients had progressive disease during the immediate post treatment period and progression occurred in less than 12 months among 50 percent of the responding patients.

Rituximab in Combination with Chemotherapy. In a Phase II study, the combination of six cycles of CHOP (cyclophosphamide, doxorubicin, vincristine, and prednisone) with rituximab given before, during, and after chemotherapy in 40 patients with predominantly untreated follicular lymphoma increased the RR (55%

CR, 40% PR), with no added related toxicity (Czuczman, Weaver, Alkuzweny, Berlfein and Grillo-Lopez, 2005). Median time to progression was 82 months.

Several randomized studies have now demonstrated that the addition of rituximab to a standard chemotherapy regimen results in higher response rates and longer time to progression, event-free survival, and overall survival for patients treated with a combination of rituximab plus chemotherapy in first line or in first relapse patients (Table 12-2). In first line patients, four studies have reported a benefit in terms of CR rates, PFS, and overall survival (Herold, et al. 2004; Hiddemann, Kneba, Dreyling, Schmitz, Lengfelder, Schmits, Reiser, Metzner, Harder, Hegewisch-Becker, Fischer, Kropff, Reis, Freund, Wormann, Fuchs, Planker, Schimke, Eimermacher, Trumper, Aldaoud, Parwaresch and Unterhalt 2005; Marcus, Imrie, Belch, Cunningham, Flores, Catalano, Solal-Celigny, Offner, Walewski, Raposo, Jack and Smith 2005; Salles, Foussard, Mounier, Morschhauser, Bosly, Lamy, Haioun, Brice, Boubadallah, Rossi, Audhuy, Fermé, Mahe, Lederlin, Sebban, Colombat and Xerri 2004). The first study randomized patients between

Table 12-2 Randomized studies comparing chemotherapy with the combination of rituximab and chemotherapy in patients with follicular lymphoma

Setting	Response rates	CR rates	Event-free survival	Time to progression
First line patients				
Marcus (Marcus, 2005) (Marcus, 2006)				
R-CVP	81%	41%	27 m	32 m
CVP	57%**	10%**	7 m**	15 m**
Hiddemann (Hiddemann, et al. 2003)				
R-CHOP	97%	20%	68 m	50 m
CHOP	90%	17%	21 m**	15 m**
Salles (Salles, 2004) (Foussard, 2006)				
R-CHVP-Ifn	Not analyzed	79%	Not reached	Not reached
CHVP-Ifn		63%**		
Herold (Herold, et al, 2004)				
R-MCP	85.5%	42%	Not reached	Not reported
MCP	65.5%**	20%**	19 m**	
Relapsing patients				
Forstpointer (Forstpointner, 2004; Forstpointner, 2006)	79%	33%	Not analyzed	16 m
R-FCM	58%**	13%**		10 m*
FCM				
Adjuvant rituximab				
Hochster (Hochster, 2004)				
CVP→R	Not reported	30%	Not reported	4.2 y
CVP		22%*		1.5 y**

* $p < 0.05$; ** $p < 0.01$

eight cycles of CVP and R-CVP (Marcus 2005). At a median follow-up of 53 months, patients treated with R-CVP had a highly significantly prolonged time to progression (median 32 months vs. 15 months for CVP; $P < 0.0001$). Median time to treatment failure was 27 months in patients receiving R-CVP and seven months in the CVP arm ($P < 0.0001$). Overall survival was longer for R-CVP than CVP (19% vs. 29% patients died, p=0.03) (Marcus, Solal-Celigny, Imrie, Catalano, Dmoszynska, Raposo, Offner and Gomez-Codina 2006). In the second study, patients were randomized between six cycles of CHOP and R-CHOP (Hiddemann 2005). In 428 patients, R-CHOP revealed a significantly higher RR (96% vs. 90%, p=0.011) and a longer TTF (median not reached vs. 2.6 years, p<0.0001).

In the French study patients were randomized between CHVP + interferon for 18 months and R-CHVP + interferon (Foussard, Mounier, Van Hoof, Delwail, Casasnovas, Deconinck, Tilly, Fitoussi, Gressin and Salles 2006). This analysis of all patients demonstrated a significant improvement of response to therapy with R-CHVP+interferon compared to CHVP+interferon, both at six months [CR+CRru 49% vs. 76%; PR 36% vs. 18%; respectively (P<.0001)], and at 18 months [CR+CRu 79% vs. 63%; PR 5% vs. 10%; respectively (P=.004)]. In the control arm, estimated 3.5 years EFS was 46% versus 67% with R-CHVP+interferon (P<.0001). Even if the median follow-up is only 3.5 years, this study showed a statistically significant overall survival advantage for patients treated with rituximab (91% compared to 84% surviving at 3.5 years, p=.029).

In relapsing patients, the FCM study showed that R-FCM is superior to FCM alone for relapsing patients with follicular or mantle cell lymphoma (Forstpointner, Dreyling, Repp, Hermann, Haenel, Metzner, Pott, Hartmann, Rothmann, Rohrberg, Boeck, Wandt, Unterhalt and Hiddemann 2004). An update of this study showed that responding patients, treated either with FCM or R-FCM, had a prolonged PFS if they received a maintenance with rituximab (Forstpointner, Unterhalt, Dreyling, Bock, Repp, Wandt, Pott, Seymour, Metzner, Hanel, Lehmann, Hartmann, Einsele and Hiddemann 2006). The EORTC study compared R-CHOP and CHOP alone in first or second line patients not previously treated with doxorubicin (van Oers, Klasa, Marcus, Wolf, Kimby, Gascoyne, Jack, Veer, Vranovsky, Holte, Glabbeke, Teodorovic, Rozewicz and Hagenbeek 2006). This last study is particularly interesting because preliminary results showed a benefit of R-CHOP over CHOP, but also a benefit of rituximab maintenance after CHOP-only induction. Rituximab maintenance yielded a median PFS from second randomization of 51.5 months versus 15 months with observation (HR, 0.40; $P < .001$). Improved PFS was found both after induction with CHOP (HR, 0.30; $P < .001$) and R-CHOP (HR, 0.54; $P = .004$). Rituximab maintenance also improved overall survival from second randomization: 85 percent at three years versus 77 percent with observation (HR, 0.52; $P =.011$).

Finally, one study reported that maintenance with rituximab in patients treated with chemotherapy increases CR rates and prolongs PFS (Hochster, Weller, Ryan, Habermann, Gascoyne, Frankel and Horning 2004). However, the role of rituximab maintenance after a combination of rituximab plus chemotherapy in first line patients remains unclear and it is not currently recommended in CR patients.

These different studies have implemented the use of combining rituximab with chemotherapy as standard treatment in patients with follicular lymphoma who need to be treated. Which of the chemotherapy regimens is better is not yet demonstrated, but the comparison of CR rates, EFS, PFS, and OS from the different studies seems to show a larger benefit with the R-CHOP regimen. The comparison of results obtained with R-CHOP to those reached with rituximab only in the same type of patients equally favors the use of R-CHOP. However, these conclusions need to be taken with caution because no randomized study has compared these different regimens.

12.7 RIT in Follicular Lymphoma

Two monoclonal antibodies have been combined with a radionucleide and have been registered for the treatment of patients with relapsing/refractory follicular lymphoma. Radioimmunotherapy with Y-90 and I-131 labeled anti-CD20 antibodies (ibritumomab tiutexan and tositumomab) was associated with a high response rate in relapsing/refractory patients. Zevalin was not tested in untreated follicular lymphoma patients.

In the initial Phase I/II study, ^{90}Y-Ibritumomab was administered on 51 patients with relapsed and refractory CD20+ B-cell non-Hodgkin's lymphoma (Witzig, White, Wiseman, Gordon, Emmanouilides, Raubitschek, Janakiraman, Gutheil, Schilder, Spies, Silverman, Parker and Grillo-Lopez 1999). The overall response rate (ORR) for the 34 patients with indolent lymphoma was 82 percent (complete response (CR) 26% and partial response (PR) 56%). The estimated median time to progression (TTP) for the entire group was 12.9+ months and the median duration of response was 11.7+ months. The major toxicity of ^{90}Y-ibritumomab was myelosuppression with thrombocytopenia being the most common. ^{90}Y-ibritumomab has been compared to rituximab in a randomized controlled Phase III study (Witzig, Gordon, Cabanillas, Czuczman, Emmanouilides, Joyce, Pohlman, Bartlett, Wiseman, Padre, Grillo-Lopez, Multany and White 2002). The ORR was 80 percent (CR /CRu 34% and PR 45%) for ^{90}Y-Ibritumomab, as compared to 56 percent (CR/CRu 20% and PR 36%) for rituximab (p=0.002). The estimated TTP was 12.6 months for ^{90}Y-Ibritumomab and 10.2 months for rituximab (p=0.062). In another study, Witzig treated with ^{90}Y-Ibritumomab 54 patients with follicular lymphoma refractory to rituximab (Witzig, Flinn, Gordon, Emmanouilides, Czuczman, Saleh, Cripe, Wiseman, Olejnik, Multani and White 2002). The ORR for the entire cohort was 74 percent (CR 14% + PR 59%). The estimated TTP and response duration for responders were 8.7 and 6.4 months, respectively.

^{131}I-Tositumomab has been studied for more than 10 years. Vose have reported the final results of a multicenter Phase II study with objectives to evaluate the efficacy, dosimetry, methodology and safety of ^{131}I-Tositumomab (Vose, Wahl, Saleh, Rohatiner, Knox, Radford, Zelenetz, Tidmarsh, Stagg and Kaminski 2000) Forty-seven patients with relapsed/refractory low-grade or transformed NHL were treated with ^{131}I-Tositumomab. The ORR for the entire group was 57 percent with 15 (32%) patients achieving CR. The ORR was similar in patients with indolent (57%) or transformed lymphoma (60%). The median duration of response was 8.2 months and 12.1 months, respectively, for each of these two groups. The treatment was well tolerated and hematologic toxicity was the principal adverse event. In the pivotal study, 60 patients with chemotherapy-refractory indolent or transformed CD20+ B-cell lymphoma (36 follicular, 23 transformed and 1 mantle cell) were treated with standard dose ^{131}I-Tositumomab (Kaminski, Zelenetz, Press, Saleh, Leonard, Fehrenbacher, Lister, Staag, Tidmarsh, Kroll, Walh, Knox and Vose 2001) The ORR was 65 percent (CR 20% and PR 45%). The median duration of response was 6.5 months. Kaminski recently presented the results of a Phase II evaluating ^{131}I-Tositumomab alone in untreated patients with follicular lymphoma (Kaminski, Tuck, Estes, Kolstad, Ross, Zasadny, Regan, Kison, Fisher, Kroll and Wahl 2005) Of the 76 patients included, more than half did not have any criteria associated with poor outcome and correspond to patients that are usually not treated. CR was observed in 75 percent of the patients, but only in 58 percent of those with a large lymph node. Median PFS was 6.1 years for all patients, but less for patients with criteria associated with poor outcome (details not given in the manuscript). This study only showed that patients without large tumors may respond well to ^{131}I-Tositumomab, but it did not allow evaluating the role of this drug in patients with follicular lymphoma.

Even though ^{131}I-Tositumomab has shown interesting results in Phase II study, duration of response is still limited. For this reason, some investigators are beginning to evaluate ^{131}I-Tositumomab in combination with other forms of therapies. In a Phase II study conducted by the SWOG (Press, Unger, Braziel, Maloney, Miller, Leblanc, Gaynor, Rivkin and Fisher 2003), ^{131}I-Tositumomab was combined with CHOP for the treatment of 90 patients with untreated follicular lymphoma. Patients received six cycles of standard CHOP followed by a consolidation dose of ^{131}I-Tositumomab if PR was achieved. The ORR after ^{131}I-Tositumomab was 90 percent (CR/CRu 67% and PR 23%). More interestingly, among patients assessable, 27 patients (57%) improved their level of response after ^{131}I-Tositumomab. The estimated two year progression-free survival (PFS) and overall survival were 81 percent and 97 percent, respectively (median follow-up, 2.3 years). SWOG is currently conducting a study comparing ^{131}I-Tositumomab and rituximab in follicular patients treated with CHOP as first treatment. Only such a study may evaluate the benefit and toxicity of Bexxar in comparison with rituximab in first line patients.

12.8 Other Monoclonal Antibodies

Several monoclonal antibodies directed against CD20 (hA20, HuMax-CD20, ocrelizumab) or other antigens (epratuzumab for CD22, apolizumab for HLA-DRB chain, galiximab for CD80) are currently in Phases I or II. No definitive conclusion can be drawn on their activity, toxicity and benefit compared to rituximab. The real interest of these new antibodies will have to be demonstrated in randomized studies.

12.9 Rituximab in Diffuse Large B-cell Lymphoma

The combination regimen R-CHOP, consisting of rituximab plus CHOP, is now considered the standard treatment for young and elderly patients with diffuse large B-cell lymphoma because of the superior activity demonstrated in three randomized studies (Table 12-3). Results from the GELA study have been recently updated with a five-year median follow-up, and showed a persisting advantage for patients treated with R-CHOP (Table 12-4) (Coiffier, Lepage, Briere, Herbrecht, Tilly, Bouabdallah, Morel, Van den Neste, Salles, Gaulard, Reyes and Gisselbrecht 2002;

Table 12-3 Randomized studies comparing CHOP with R-CHOP in patients with diffuse large B-cell lymphoma

	Coiffier (Coiffier, 2002; Feugier, 2005)	Habermann (Habermann, 2006)	Pfreundschuh (Pfreundschuh, 2006)
Setting	60–80 years old No stage I CHOP	60–80 years old No stage I Maintenance	<60 years old IPI 0-1 CHOP or CHOP-like
Median follow-up	5 years	2.7 years	2 years
CR rates			
R-CHOP	75%	78%	85%
CHOP	63%**	77%	65%**
Early progression rates			
R-CHOP	9%	15%	16%
CHOP	22%**	17%	5%**
Relapses			
R-CHOP	34%	Not reported	Not reported
CHOP	20%**		
Event-free survival			2-year TTF
R-CHOP	3.8 y	3.4 y	81%
CHOP	1.1 y**	2.4 y	58%**
Progression-free survival			
R-CHOP	Not reported	Not reported	Not reached
CHOP	1.0 y**		
Overall survival			2-year OS
R-CHOP	Not reached	Not different	95%
CHOP	3.1y**		85%**

* $p < 0.05$, ** $p < 0.01$

Table 12-4 Five-year survivals observed in the GELA study comparing eight cycles of R-CHOP and CHOP in elderly patients with diffuse large B-cell lymphoma (Feugier 2005)

	R-CHOP	CHOP	P value
Median event-free survival	3.8 y	1.1 y	= 0.00002
5-year event-free survival	47%	29%	
Median progression-free survival	Not reached	1 y	< 0.00001
5-year progression-free survival	54%	30%	
Median overall survival	Not reached	3.1 y	= 0.0073
5-year overall survival	58%	45%	

Feugier, Van Hoof, Sebban, Solal-Celigny, Bouabdallah, Ferme, Christian, Lepage, Tilly, Morschhauser, Gaulard, Salles, Bosly, Gisselbrecht, Reyes and Coiffier 2005). In this study, patients with DLBCL and aged 60 to 80 years were treated either with eight cycles of CHOP or eight cycles of CHOP, combined with rituximab (R-CHOP). The difference observed between the two arms was already statistically significant for EFS, PFS, and OS with a median follow-up of one year and improved with follow-up.

The MInT study compared in 824 patients six cycles of R-CHOP-like chemotherapy to CHOP-like in young patients with a low-risk DLBCL (Pfreundschuh, Trumper, Osterborg, Pettengell, Trneny, Imrie, Ma, Gill, Walewski, Zinzani, Stahel, Kvaloy, Shpilberg, Jaeger, Hansen, Lehtinen, Lopez-Guillermo, Corrado, Scheliga, Milpied, Mendila, Rashford, Kuhnt and Loeffler 2006). After a median time 34 months follow-up, R-CHEMO patients had a significantly longer TTF ($p<0.00001$), with estimated two-year TTF rates of 60 percent (CHEMO) vs. 76 percent (R-CHEMO). Similarly, overall survival was significantly different ($p<0.001$), with two-year survival rates of 87 percent (CHEMO) and 94 percent (R-CHEMO), respectively. The American study (ECOG/SWOG/CALGB study) was associated with a statistical benefit in the primary endpoint time to treatment failure (TTF) for the addition of rituximab to CHOP versus CHOP alone (Habermann, Weller, Morrison, Gascoyne, Cassileth, Cohn, Dakhil, Woda, Fisher, Peterson and Horning 2006). However, the complicated design of this study makes conclusions difficult compared to the two other studies. The interesting point of this study is the second randomization looking at the effect of rituximab in patients who reached a CR or a PR. If rituximab maintenance may decrease the progression rate in patients treated with CHOP only, it has no effect on patients treated with R-CHOP.

12.10 Monoclonal Antibodies in Other Lymphomas

Small Lymphocytic Lymphoma. The efficacy of rituximab alone in this lymphoma is not very well known, with few and discordant results. In a European study in relapsing patients, the efficacy was low, with only a 10 percent RR (Foran, Rohatiner, Cunningham, Popescu, Solal-Celigny, Ghielmini, Coiffier, Johnson, Gisselbrecht, Reyes, Radford, Bessell, Souleau, Benzohra and Lister 2000). In untreated patients, in contrast, Hainsworth found a 51 percent RR after four injections, with only 4 percent CR, and a median progression-free survival of 18.6 months (Hainsworth, Litchy, Barton, Houston, Hermann, Bradof and Greco 2003).

Marginal Zone Lymphoma. Mostly case reports have shown an efficacy of rituximab in these lymphomas. Efficacy was demonstrated in relapsing mucosa-associated lymphoid tissue (MALT) lymphoma (Conconi, Martinelli, Thieblemont, Ferreri, Devizzi, Peccatori, Ponzoni, Pedrinis, Dell'Oro, Pruneri, Filipazzi, Dietrich, Gianni, Coiffier, Cavalli and Zucca 2003). A current IELSG (International Extranodal Lymphoma Study Group) trial randomizes chlorambucil vs. chlorambucil + rituximab in new or relapsing patients with MALT lymphoma.

Mantle Cell Lymphoma. Mantle cell lymphoma (MCL) has indolent lymphoma characteristics, but tends to pursue an aggressive clinical course and is incurable with standard chemotherapy. An interim analysis of a randomized trial comparing FCM (fludarabine/cyclophosphamide/mitoxantrone) to FCM plus rituximab has shown a striking improvement in RR with rituximab (65% vs. 33%; CR 35% vs.0%), with a trend towards longer overall survival (Forstpointner, 2004). Interestingly, about one-third of the patients achieved a molecular remission. A maintenance with rituximab was scheduled for responding patients and enabled a prolonged duration of response (Forstpointner 2006). Long-term remissions have been reported with intensive chemotherapy and autologous stem cell transplantation plus rituximab (see 12.11).

Chronic Lymphocytic Leukemia (CLL). Rituximab, given weekly as a single agent, has low activity in relapsing patients with CLL. A better activity has been observed in untreated patients (Hainsworth, 2003). Dose escalation, achieved by a thrice-weekly dosing schedule (Byrd, Murphy, Howard, Lucas, Goodrich, Park, Pearson, Waselenko, Ling, Grever, Grillo-Lopez, Rosenberg, Kunkel and Flinn 2001), or higher weekly doses – 500 to 2250 mg/m^2 (O'Brien, Kantarjian, Thomas, Giles, Freireich, Cortes, Lerner and Keating 2001) – is necessary to reach significant clinical activity, with a RR of respectively 45 percent and 36 percent ., as a single agent. The concurrent administration of rituximab with fludarabine resulted in better results with a RR rate of 90 percent, with 47 percent CR (Byrd, Peterson, Morrison, Park, Jacobson, Hoke, Vardiman, Rai, Schiffer and Larson 2003). Ongoing clinical studies are examining the use of rituximab associated with fludarabine and cyclophosphamide, which has shown great promise in a single-center Phase II study (Keating, O'Brien, Albitar, Lerner, Plunkett, Giles, Andreeff, Cortes, Faderl, Thomas, Koller, Wierda, Detry, Lynn and Kantarjian 2005; Wierda, O'Brien, Wen, Faderl, Garcia-Manero, Thomas, Do, Cortes, Koller, Beran, Ferrajoli, Giles, Lerner, Albitar, Kantarjian and Keating 2005).

Alemtuzumab is a humanized monoclonal antibody active against CD52. Compared to CD20, CD52 is expressed at much higher density on the surface of CLL cells. Activity of alemtuzumab in fludarabine-refractory CLL was established in the pivotal trial conducted by Keating (Keating, Flinn, Jain, Binet, Hillmen, Byrd, Albitar, Brettman, Santabarbara, Wacker and Rai 2002). Among 93 patients, the overall response rate was 33 percent, including 2 percent complete responders. The median time to response was six weeks and the median time of therapy extended up to eight weeks. Median survival of all patients was 16 months, but was 32 months for the responding patients.

Other Lymphomas. In post-transplant lymphoproliferative disease several Phase II patients have shown a good activity with rituximab alone or in combination with chemotherapy (Blaes, Peterson, Bartlett, Dunn and Morrison, 2005; Jain, Marcos, Pokharna, Shapiro, Fontes, Marsh, Mohanka and Fung 2005; Milpied, Vasseur, Parquet, Garnier, Antoine, Quartier, Carret, Bouscary, Faye, Bourbigot, Reguerre, Stoppa, Bourquard, de Ligny, Dubief, Mathieu-Boue and Leblond 2000).

The only yet reported randomized study without benefit in the rituximab arm was in patients with HIV-associated lymphoma; the response rate was not statistically different in R-CHOP or CHOP arms (Kaplan, Scadden and for the AIDS Malignancies Consortium 2004). However, this study did not have the power to show a significant difference and the follow-up is extremely short. In another Phase II study, R-CHOP produced a CR rate of 77 percent and a two-year survival rate of 75 percent without severe infectious complications (Boue, Gabarre, Gisselbrecht, Reynes, Cheret, Bonnet, Billaud, Raphael, Lancar and Costagliola 2006).

Alemtuzumab have been reported active in cutaneous T-cell lymphomas and peripheral T-cell lymphoma, but activity was low and of short duration in most of the cases (Enblad, Hagberg, Erlanson, Lundin, MacDonald, Repp, Schetelig, Seipelt, Osterborg, Lundin, Kimby, Bjorkholm, Broliden, Celsing, Hjalmar, Mollgard, Rebello, Hale, Waldmann, Mellstedt and Osterborg 2004; Lundin, Hagberg, Repp, Cavallin-Stahl, Freden, Juliusson, Rosenblad, Tjonnfjord, Wiklund and Osterborg 2003).

The vast majority of the immunotoxin trials have been Phase I studies designated to determine the maximum tolerated dose. These trials have shown that therapeutic serum levels may be achieved with tolerable toxicity. A relatively uniform toxicity has been observed with vascular leak syndrome, hepatotoxicity, and myalgia. The different trials have shown a low response rate of 10 percent to 25 percent partial responses without durable efficacy. The only available immunotoxin for treating patients is denileukin diftitox (OnTak®) for the treatment of cutaneous T-cell lymphoma: 30 percent of them responded with 10 percent of complete responses (Olsen, Duvic, Frankel, Kim, Martin, Vonderheid, Jegasothy, Wood, Gordon, Heald, Oseroff, Pinter-Brown, Bowen, Kuzel, Fivenson, Foss, Glode, Molina, Knobler, Stewart, Cooper, Stevens, Craig, Reuben, Bacha, et al. 2001). The future of this therapy will depend on decreasing toxicity, decreasing immune response against the construct, and on increasing the antitumor activity.

12.11 MAb and High Dose Therapy (HDT) with Autologous Stem Cell Transplantation (ASCT)

Rituximab has been used as an in vivo purging agent before and as maintenance therapy after ASCT, in follicular and mantle cell lymphoma (Belhadj, Delfau-Larue, Elgnaoui, Beaujean, Beaumont, Pautas, Gaillard, Kirova, Allain, Gaulard, Farcet, Reyes and Haioun, 2004; Gianni, Magni, Martelli, Di Nicola, Carlo-Stella, Pilotti, Rambaldi, Cortelazzo, Patti, Parvis, Benedetti, Capria, Corradini, Tarella and Barbui 2003), and in aggressive lymphoma (Horwitz and Horning 2004) – in first line or in relapse – with promising results. An ongoing international trial in relapsed and refractory aggressive lymphoma randomizes rituximab-DHAP (dexamethasone, aracytine, cisplatin) vs. rituximab-ICE (ifosfamide, carboplatin, etoposide) before ASCT and with a second randomization between rituximab maintenance and observation (CORAL study) (Hagberg and Gisselbrecht 2006).

Rituximab given after ASCT might have the interest to complete the remission and to further decrease the relapse rate. However, this treatment may be associated with more infections.(Neumann, Harmsen, Martin, Kronenwett, Kondakci, Aivado, Germing, Haas and Kobbe 2006) It had been associated with a severe decrease in immunoglobulin levels (Lim, Zhang, Wang, Esler, Beggs, Pruitt, Hancock and Townsend 2005) and more frequent neutropenia (Lemieux 2004).

A few studies have looked at the potential use of radiolabeled monoclonal antibodies in the context of HDT. As compared to external TBI, radiolabeled monoclonal antibodies could, theoretically, permit delivering a higher dose of radiation to the tumor while limiting radiation dose to normal tissues, thus potentially reducing toxicity and treatment–related mortality (Press, Eary, Appelbaum, Martin, Nelp, Glenn, Fisher, Porter, Matthews, Gooley and Bernstein 1995). Press used myeloablative doses of ^{131}I-Tositumomab in combination with chemotherapy (Press, Eary, Gooley, Gopal, Liu, Rajendran, Maloney, Petersdorf, Bush, Durack, Martin, Fisher, Wood, Borrow, Porter, Smith, Matthews, Appelbaum and Bernstein 2000). In this Phase I/II study, 25 Gy was considered the maximum dose of radiation that could be delivered to critical normal organs when combined with cyclophosphamide (100 mg/kg) and etoposide (60 mg/kg). They observed an objective response of 87 percent in a population of the patients with relapsed B-cell lymphoma (73% indolent lymphoma). The estimated two years OS and PFS were 83 percent and 68 percent.

12.12 Conclusion

Rituximab was the first monoclonal antibody registered in the treatment of lymphomas and it has allowed one of the major progresses for the treatment of lymphoma patients. Alone, it is a very well tolerated drug and it has a great activity in relapsing patients. However, it will hardly result in a cure in this setting. In combination with chemotherapy, rituximab allowed for the highest response rates and longest event-free and overall survivals ever described in follicular and diffuse large B-cell lymphomas. It has activity, but less well demonstrated in other B-cell lymphomas. Other monoclonal antibodies targeting CD20 or other antigens are on their way, but their activity is not yet well defined compared to rituximab. Radioimmunotherapy may add some specific activity, but here too this is not well demonstrated. Antibodies conjugated with toxin are less used for the moment.

References

Alas, S., Bonavida, B. Rituximab inactivates signal transducer and activation of transcription 3 (STAT3) activity in B-non-Hodgkin's lymphoma through inhibition of the interleukin 10 autocrine/paracrine loop and results in down-regulation of Bcl-2 and sensitization to cytotoxic drugs. Cancer Res. 2001;61:5137–44.

Alas, S., Bonavida, B., Emmanouilides, C. Potentiation of fludarabine cytotoxicity on non-Hodgkin's lymphoma by pentoxifylline and Rituximab. Anticancer Res. 2000;20:2961–6.

Ardeshna, K. M., Smith, P., Norton, A., Hancock, B. W., Hoskin, P. J., MacLennan, K. A., Marcus, R. E., Jelliffe, A., Hudson, G. V., Linch, D. C. Long-term effect of a watch and wait policy versus immediate systemic treatment for asymptomatic advanced-stage non-Hodgkin's lymphoma: a randomised controlled trial. Lancet. 2003;362:516–22.

Belhadj, K., Delfau-Larue, M. H., Elgnaoui, T., Beaujean, F., Beaumont, J. L., Pautas, C., Gaillard, I., Kirova, Y., Allain, A., Gaulard, P., Farcet, J. P., Reyes, F., Haioun, C. Efficiency of in vivo purging with rituximab prior to autologous peripheral blood progenitor cell transplantation in B-cell non-Hodgkin's lymphoma: a single institution study. Ann Oncol. 2004;15:504–10.

Berinstein, N. L., Grillolopez, A. J., White, C. A., Bencebruckler, I., Maloney, D., Czuczman, M., Green, D., Rosenberg, J., McLaughlin, P., Shen, D. Association of serum rituximab (IDEC-C2B8) concentration and antitumor response in the treatment of recurrent low-grade or follicular non-Hodgkin's lymphoma Ann Oncol. 1998;9:995–1001.

Bienvenu, J., Chvetzoff, R., Salles, G., Balter, C., Tilly, H., Herbrecht, R., Morel, P., Lederlin, P., Solal-Celigny, P., Audhuy, B., Christian, B., Gabarre, J., Casasnovas, O., Marit, G., Sebban, C., Coiffier, B. Tumor necrosis factor alpha release is a major biological event associated with rituximab treatment. Hematol J. 2001;2:378–84.

Blaes, A. H., Peterson, B. A., Bartlett, N., Dunn, D. L., Morrison, V. A. Rituximab therapy is effective for post-transplant lymphoproliferative disorders after solid organ transplantation: results of a phase II trial. Cancer. 2005;104:1661–7.

Boettcher, S., Pott, C., Ritgen, M., Hiddemann, W., Unterhalt, M., Kneba, M. Evidence for Fcγ receptor IIIA-independent rituximab effector mechanisms in patients with follicular lymphoma treated with combined immuno-chemotherapy. In: Blood, editor. ASH 2004; 2004: Blood; 2004. p. 170a.

Bohen, S. P., Troyanskaya, O. G., Alter, O., Warnke, R., Botstein, D., Brown, P. O., Levy, R. Variation in gene expression patterns in follicular lymphoma and the response to rituximab. Proc Nat Acad Sci USA. 2003;100:1926–30.

Boue, F., Gabarre, J., Gisselbrecht, C., Reynes, J., Cheret, A., Bonnet, F., Billaud, E., Raphael, M., Lancar, R., Costagliola, D. Phase II trial of CHOP plus rituximab in patients with HIV-associated non-Hodgkin's lymphoma. J Clin Oncol. 2006;24:4123–8.

Byrd, J. C., Murphy, T., Howard, R. S., Lucas, M. S., Goodrich, A., Park, K., Pearson, M., Waselenko, J. K., Ling, G., Grever, M. R., Grillo-Lopez, A. J., Rosenberg, J., Kunkel, L., Flinn, I. W. Rituximab using a thrice-weekly dosing schedule in B-Cell chronic lymphocytic leukemia and small lymphocytic lymphoma demonstrates clinical activity and acceptable toxicity. J Clin Oncol. 2001;19:2153–64.

Byrd, J. C., Peterson, B. L., Morrison, V. A., Park, K., Jacobson, R., Hoke, E., Vardiman, J. W., Rai, K., Schiffer, C. A., Larson, R. A. Randomized phase II study of fludarabine with concurrent versus sequential treatment with rituximab in symptomatic, untreated patients with B-cell chronic lymphocytic leukemia: results from Cancer and Leukemia Group B 9712 (CALGB 9712). Blood. 2003;101:6–14.

Cartron, G., Dacheux, L., Salles, G., Solal-Celigny, P., Bardos, P., Colombat, P., Watier, H. Therapeutic activity of humanized anti-CD20 monoclonal antibody and polymorphism in IgG Fc receptor Fc gamma RIIIa gene. Blood. 2002;99:754–8.

Cartron, G., Watier, H., Golay, J., Solal-Celigny, P. From the bench to the bedside: ways to improve rituximab efficacy. Blood. 2004;104:2635–42.

Cheson, B. D. Radioimmunotherapy of non-Hodgkin's Lymphomas. Blood. 2003;101:391–8.

Coiffier, B. Current strategies for the treatment of diffuse large B-cell lymphoma. Cur Op Hematol. 2005;12:In press.

Coiffier, B. First-line treatment of follicular lymphoma in the era of monoclonal antibodies. Clin Adv Hematol Oncol. 2005;3:484–91.

Coiffier, B., Lepage, E., Briere, J., Herbrecht, R., Tilly, H., Bouabdallah, R., Morel, P., Van den Neste, E., Salles, G., Gaulard, P., Reyes, F., Gisselbrecht, C. CHOP chemotherapy plus rituximab compared with CHOP alone in elderly patients with diffuse large-B-cell lymphoma. N Engl J Med. 2002;346:235–42.

Colombat, P., Salles, G., Brousse, N., Eftekhari, P., Soubeyran, P., Delwail, V., Deconinck, E., Haioun, C., Foussard, C., Sebban, C., Stamatoullas, A., Milpied, N., Boue, F., Taillan, B., Lederlin, P., Najman, A., Thieblemont, C., Montestruc, F., Mathieu-Boue, A., Benzohra, A., Solal-Celigny, P. Rituximab (anti-CD20 monoclonal antibody) as single first-line therapy for patients with follicular lymphoma with a low tumor burden: clinical and molecular evaluation. Blood. 2001;97:101–6.

Conconi, A., Martinelli, G., Thieblemont, C., Ferreri, A. J. M., Devizzi, L., Peccatori, F., Ponzoni, M., Pedrinis, E., Dell'Oro, S., Pruneri, G., Filipazzi, V., Dietrich, P. Y., Gianni, A. M., Coiffier, B., Cavalli, F., Zucca, E. Clinical activity of rituximab in extranodal marginal zone B-cell lymphoma of MALT type. Blood. 2003;102:2741–5.

Czuczman, M. S., Weaver, R., Alkuzweny, B., Berlfein, J., Grillo-Lopez, A. J. Prolonged Clinical and Molecular Remission in Patients With Low-Grade or Follicular Non-Hodgkin's Lymphoma Treated With Rituximab Plus CHOP Chemotherapy: 9-Year Follow-Up. J Clin Oncol. 2005;23:DOI:10.1200/JCO.2004.04.020.

Davis, T. A., Grillo-Lopez, A. J., White, C. A., McLaughlin, P., Czuczman, M. S., Link, B. K., Maloney, D. G., Weaver, R. L., Rosenberg, J., Levy, R. Rituximab anti-CD20 monoclonal antibody therapy in non-Hodgkin's lymphoma: Safety and efficacy of re-treatment. J Clin Oncol. 2000;18:3135–43.

Demidem, A., Lam, T., Alas, S., Hariharan, K., Hanna, N., Bonavida, B. Chimeric Anti-Cd20 (Idec-C2b8) Monoclonal Antibody Sensitizes a B-Cell Lymphoma Cell Line to Cell Killing By Cytotoxic Drugs. Cancer Bioth Radiopharm. 1997;12:177–86.

Dillman, R. O. Radiolabeled anti-CD20 monoclonal antibodies for the treatment of B-cell Lymphoma. Journal of clinical oncology. 2002;20:3545–57.

Enblad, G., Hagberg, H., Erlanson, M., Lundin, J., MacDonald, A. P., Repp, R., Schetelig, J., Seipelt, G., Osterborg, A., Lundin, J., Kimby, E., Bjorkholm, M., Broliden, P. A., Celsing, F., Hjalmar, V., Mollgard, L., Rebello, P., Hale, G., Waldmann, H., Mellstedt, H., Osterborg, A. A pilot study of alemtuzumab (anti-CD52 monoclonal antibody) therapy for patients with relapsed or chemotherapy-refractory peripheral T-cell lymphomas.

Phase II trial of subcutaneous anti-CD52 monoclonal antibody alemtuzumab (Campath-1H) as first-line treatment for patients with B-cell chronic lymphocytic leukemia (B-CLL). Blood. 2004;103:2920–4.

Feugier, P., Van Hoof, A., Sebban, C., Solal-Celigny, P., Bouabdallah, R., Ferme, C., Christian, B., Lepage, E., Tilly, H., Morschhauser, F., Gaulard, P., Salles, G., Bosly, A., Gisselbrecht, C., Reyes, F., Coiffier, B. Long-term results of the R-CHOP study in the treatment of elderly patients with diffuse large B-cell lymphoma: a study by the Groupe d'Etude des Lymphomes de l'Adulte. J Clin Oncol. 2005;23:4117–26.

Foran, J. M., Rohatiner, A. Z. S., Cunningham, D., Popescu, R. A., Solal-Celigny, P., Ghielmini, M., Coiffier, B., Johnson, P. W. M., Gisselbrecht, C., Reyes, F., Radford, J. A., Bessell, E. M., Souleau, B., Benzohra, A., Lister, T. A. European phase II study of rituximab (chimeric anti-CD20 monoclonal antibody) for patients with newly diagnosed mantle-cell lymphoma and previously treated mantle-cell lymphoma, immunocytoma, and small B-cell lymphocytic lymphoma. J Clin Oncol. 2000;18:317–24.

Forstpointner, R., Dreyling, M., Repp, R., Hermann, S., Haenel, A., Metzner, B., Pott, C., Hartmann, F., Rothmann, F., Rohrberg, R., Boeck, H. P., Wandt, H., Unterhalt, M., Hiddemann, W. The addition of rituximab to a combination of fludarabine, cyclophosphamide, mitoxantrone (FCM) significantly increases the response rate and prolongs survival as compared to FCM alone in patients with relapsed and refractory follicular and mantle cell lymphomas - results of a prospective randomized study of the German low grade lymphoma study group (GLSG). Blood. 2004;104:3064–71.

Forstpointner, R., Unterhalt, M., Dreyling, M., Bock, H. P., Repp, R., Wandt, H., Pott, C., Seymour, J. F., Metzner, B., Hanel, A., Lehmann, T., Hartmann, F., Einsele, H., Hiddemann, W. Maintenance therapy with rituximab leads to a significant prolongation of response duration after salvage therapy with a combination of rituximab, fludarabine, cyclophosphamide and mitoxantrone (R-FCM) in patients with relapsed and refractory follicular and mantle cell lymphomas - results of a prospective randomized study of the German low grade lymphoma study group (GLSG). Blood. 2006;DOI: 10.1182/blood-2006–04–016725.

Foussard, C., Mounier, N., Van Hoof, A., Delwail, V., Casasnovas, O., Deconinck, E., Tilly, H., Fitoussi, O., Gressin, R., Salles, G. Update of the FL2000 randomized trial combining rituximab to CHVP-Interferon in follicular lymphoma (FL) patients (pts). J Clin Oncol. 2006;24.

Ghetie, M. A., Bright, H., Vitetta, E. S. Homodimers. but not monomers of Rituxan (chimeric anti-CD20) induce apoptosis in human B-lymphoma cells and synergize with a chemotherapeutic agent and an immunotoxin. Blood. 2001;97:1392–8.

Ghetie, M. A., Crank, M., Kufert, S., Pop, I., Vitetta, E. Rituximab, but not other anti-CD20 antibodies reverses multidrug resistance in 2 B lymphoma cell lines, blocks the activity of P-glycoprotein (P-gp), and induces P-gp to translocate out of lipid rafts. J Immunother. 2006;29:536–44.

Ghielmini, M., Schmitz, S. F., Cogliatti, S. B., Pichert, G., Hummerjohann, J., Waltzer, U., Fey, M. F., Betticher, D. C., Martinelli, G., Peccatori, F., Hess, U., Zucca, E., Stupp, R., Kovacsovics, T., Helg, C., Lohri, A., Bargetzi, M., Vorobiof, D., Cerny, T. Prolonged treatment with rituximab in patients with follicular lymphoma significantly increases event-free survival and response duration compared with the standard weekly x 4 schedule. Blood. 2004;103:4416–23.

Gianni, A. M., Magni, M., Martelli, M., Di Nicola, M., Carlo-Stella, C., Pilotti, S., Rambaldi, A., Cortelazzo, S., Patti, C., Parvis, G., Benedetti, F., Capria, S., Corradini, P., Tarella, C., Barbui, T. Long-term remission in mantle cell lymphoma following high-dose sequential chemotherapy and in vivo rituximab-purged stem cell autografting (R-HDS regimen). Blood. 2003;102:749–55.

Grillo-Lopez, A. J., Hedrick, E., Rashford, M., Benyunes, M. Rituximab: Ongoing and future clinical development. Sem Oncol. 2002;29:105–12.

Habermann, T. M., Weller, E. A., Morrison, V. A., Gascoyne, R. D., Cassileth, P. A., Cohn, J. B., Dakhil, S. R., Woda, B., Fisher, R. I., Peterson, B. A., Horning, S. J. Rituximab-CHOP versus CHOP alone or with maintenance rituximab in older patients with diffuse large B-cell lymphoma. J Clin Oncol. 2006;24:3121–7.

Hagberg, H., Gisselbrecht, C. Randomized phase III study of R-ICE versus R-DHAP in relapsed patients with CD20 diffuse large B-cell lymphoma (DLBCL) followed by high-dose therapy and a second randomization to maintenance treatment with rituximab or not: an update of the CORAL study. Ann Oncol. 2006;17 Suppl 4:iv31–2.

Hainsworth, J. D., Litchy, S., Barton, J. H., Houston, G. A., Hermann, R. C., Bradof, J. E., Greco, F. A. Single-agent rituximab as first-line and maintenance treatment for patients with chronic lymphocytic leukemia or small lymphocytic lymphoma: A phase II trial of the Minnie Pearl Cancer Research Network. J Clin Oncol. 2003;21:1746–51.

Hainsworth, J. D., Litchy, S., Burris, H. A., Scullin, D. C., Corso, S. W., Yardley, D. A., Morrissey, L., Greco, F. A. Rituximab as first-line and maintenance therapy for patients with indolent non-Hodgkin's lymphoma. J Clin Oncol. 2002;20:4261–7.

Hainsworth, J. D., Litchy, S., Shaffer, D. W., Lackey, V. L., Grimaldi, M., Greco, F. A. Maximizing therapeutic benefit of rituximab: maintenance therapy versus re-treatment at progression in patients with indolent non-Hodgkin's lymphoma–a randomized phase II trial of the Minnie Pearl Cancer Research Network. J Clin Oncol. 2005;23:1088–95.

Herold, M., et al. Results of a prospective randomized open label phase III study comparing rituximab plus mitoxantrone, chlorambucil, prednisolone chemotherapy (R-MCP) versus MCP alone in untreated advanced indolent non-Hodgkin's lymphoma and mantle cell lymphoma. In: Blood, editor. ASH; 2004; San Diego: Blood; 2004. p. 169a.

Hiddemann, W., Dreyling, M. H., Forstpointner, R., Kneba, M., Woermann, B., Lengfelder, E., Schmits, R., Reiser, M., Metzner, B., Schmitz, N., Truemper, L., Eimermacher, H., Parwaresch, R. Combined Immuno-Chemotherapy (R-CHOP) Significantly Improves Time To Treatment Failure in First Line Therapy of Follicular Lymphoma Results of a Prospective Randomized Trial of the German Low Grade Lymphoma Study Group (GLSG). In: Blood, editor. American Society of Hematology; 2003; San Diego, CA: Blood; 2003. p. 104a.

Hiddemann, W., Kneba, M., Dreyling, M., Schmitz, N., Lengfelder, E., Schmits, R., Reiser, M., Metzner, B., Harder, H., Hegewisch-Becker, S., Fischer, T., Kropff, M., Reis, H. E., Freund, M., Wormann, B., Fuchs, R., Planker, M., Schimke, J., Eimermacher, H., Trumper, L., Aldaoud, A., Parwaresch, R., Unterhalt, M. Frontline therapy with rituximab added to the combination of cyclophosphamide, doxorubicin, vincristine, and prednisone (CHOP) significantly improves the outcome for patients with advanced-stage follicular lymphoma compared with therapy with CHOP alone: results of a prospective randomized study of the German Low-Grade Lymphoma Study Group. Blood. 2005;106:3725–32.

Hochster, H. S., Weller, E., Ryan, T., Habermann, T. M., Gascoyne, R., Frankel, S. R., Horning, S. J. Results of E1496: A phase III trial of CVP with or without maintenance with rituximab in advanced indolent lymphoma. In: ASCO P, editor. ASCO; 2004; New Orleans, LA: Proc ASCO; 2004. p. 556.

Horwitz, S. M., Horning, S. J. Rituximab in stem cell transplantation for aggressive lymphoma. Curr Hematol Rep. 2004;3:227–9.

Jain, A. B., Marcos, A., Pokharna, R., Shapiro, R., Fontes, P. A., Marsh, W., Mohanka, R., Fung, J. J. Rituximab (Chimeric Anti-CD20 Antibody) for Posttransplant Lymphoproliferative Disorder after Solid Organ Transplantation in Adults: Long-Term Experience from a Single Center. Transplantation. 2005;80:1692–8.

Janas, E., Priest, R., Wilde, J. I., White, J. H., Malhotra, R. Rituxan (anti-CD20 antibody)-induced translocation of CD20 into lipid rafts is crucial for calcium influx and apoptosis. Clin Exp Immunol. 2005;139:439–46.

Juweid, M. E. Radioimmunotherapy of B-Cell Non-Hodgkin's lymphoma: From clinical trials to clinical practice. Journal of Nuclear Medecine. 2002;43:1507–29.

Kaminski, M. S., Tuck, M., Estes, J., Kolstad, A., Ross, C. W., Zasadny, K., Regan, D., Kison, P., Fisher, S., Kroll, S., Wahl, R. L. 131I-tositumomab therapy as initial treatment for follicular lymphoma. N Engl J Med. 2005;352:441–9.

Kaminski, M. S., Zelenetz, A. D., Press, O. W., Saleh, M., Leonard, J., Fehrenbacher, L., Lister, A. T., Staag, R. J., Tidmarsh, G. F., Kroll, S., Walh, R. L., Knox, S. J., Vose, J. M. Pivotal study of iodine I 131 tositumomab for chemotherapy-refractory low-grade or transformed low-grade B-cell non-Hodgkin's lymphoma. Journal of clinical oncology. 2001;19:3918–28.

Kaplan, L. D., Scadden, D. T. for the AIDS Malignancies Consortium. No benefit from rituximab in a randomized phase III trial of CHOP with or without rituximab for patients with HIV-associated non-Hodgkin's lymphoma: AIDS malignancies consortium study 010. In: Oncol PASC, editor. ASCO; 2004: Proc Am Soc Clin Oncol 2004. p. 564.

Keating, M. J., Flinn, I., Jain, V., Binet, J. L., Hillmen, P., Byrd, J., Albitar, M., Brettman, L., Santabarbara, P., Wacker, B., Rai, K. R. Therapeutic role of alemtuzumab (Campath-1H) in patients who have failed fludarabine: results of a large international study. Blood. 2002;99:3554–61.

Keating, M. J., O'Brien, S., Albitar, M., Lerner, S., Plunkett, W., Giles, F., Andreeff, M., Cortes, J., Faderl, S., Thomas, D., Koller, C., Wierda, W., Detry, M. A., Lynn, A., Kantarjian, H. Early results of a chemoimmunotherapy regimen of fludarabine, cyclophosphamide, and rituximab as initial therapy for chronic lymphocytic leukemia. J Clin Oncol. 2005;23:4079–88.

Kennedy, A. D., Beum, P. V., Solga, M. D., DiLillo, D. J., Lindorfer, M. A., Hess, C. E., Densmore, J. J., Williams, M. E., Taylor, R. P. Rituximab infusion promotes rapid complement depletion and acute CD20 loss in chronic lymphocytic leukemia. J Immunol. 2004;172:3280–8.

Kim, D. H., Jung, H. D., Kim, J. G., Lee, J. J., Yang, D. H., Park, Y. H., Do, Y. R., Shin, H. J., Kim, M. K., Hyun, M. S., Sohn, S. K. FcGRIIIa gene polymorphisms may correlate with response to frontline R-CHOP therapy for diffuse large B-cell lymphoma. Blood. 2006;108:2720–5.

Kimby, E. Tolerability and safety of rituximab (MabThera). Cancer Treat Rev. 2005;31:456–73.

Lemieux, B., Bouafia, F., Thieblemont, C., Hequet, O., Arnaud, P., Tartas, S., Traulle, C., Salles, G., Coiffier, B. Second treatment with rituximab in B-cell non-Hodgkin's lymphoma: efficacy and toxicity on 41 patients treated at CHU-Lyon Sud. Hematol J. 2004;In press.

Lemieux, B., Coiffier, B. Radio-immunotherapy in low-grade non-Hodgkin's lymphoma Best Practice & Research Clinical Haematology 2005;18:81–95.

Lemieux, B., Tartas, S., Traulle, C., Espinouse, D., Thieblemont, C., Bouafia, F., Alhusein, Q., Antal D., Salles, G., Coiffier, B. Rituximab-related late-onset neutropenia after autologous stem cell transplantation for aggressive non-Hodgkin's lymphoma. Bone Marrow Transplant. 2004;33:921–3.

Lim, S. H., Zhang, Y., Wang, Z., Esler, W. V., Beggs, D., Pruitt. B., Hancock, P., Townsend, M. Maintenance rituximab after autologous stem cell transplant for high-risk B-cell lymphoma induces prolonged and severe hypogammaglobulinemia. Bone Marrow Transplant. 2005;35:207–8.

Lim, S. H., Zhang, Y., Wang, Z., Esler, W. V., Beggs, D., Pruitt, B., Hancock, P., Townsend, M. Maintenance rituximab after autologous stem cell transplant for high-risk B-cell lymphoma induces prolonged and severe hypogammaglobulinemia. 46th meeting of the American Socity of Hematology; 2004; San Diego, CA: Blood; 2004. p. 395a.

Lundin, J., Hagberg, H., Repp, R., Cavallin-Stahl, E., Freden, S., Juliusson, G., Rosenblad, E., Tjonnfjord, G., Wiklund, T., Osterborg, A. Phase 2 study of alemtuzumab (anti-CD52 monoclonal antibody) in patients with advanced mycosis fungoides/Sezary syndrome. Blood. 2003;101:4267–72.

Maloney, D. G. Preclinical and phase I and II trials of rituximab. Sem Oncol. 1999;26:74–8.

Marcus, R., Imrie, K., Belch, A., Cunningham, D, Flores, E., Catalano, J., Solal-Celigny, P., Offner, F., Walewski, J., Raposo, J., Jack, A., Smith, P. CVP chemotherapy plus rituximab compared with CVP as first-line treatment for advanced follicular lymphoma. Blood. 2005;105:1417–23.

Marcus, R., Solal-Celigny, P., Imrie, K., Catalano, J., Dmoszynska, A., Raposo, J., Offner, F., Gomez-Codina, J. MabThera (Rituximab) Plus Cyclophosphamide, Vincristine and Prednisone (CVP) Chemotherapy Improves Survival in Previously Untreated Patients with Advanced Follicular Non-Hodgkins Lymphoma (NHL). Blood. 2006;108:

McLaughlin, P., Grillolopez, A. J., Link, B. K., Levy, R., Czuczman, M. S., Williams, M. E., Heyman, M. R., Bencebruckler, I., White, C. A., Cabanillas, F., Jain, V., Ho, A. D., Lister, J., Wey, K., Shen, D., Dallaire, B. K. Rituximab Chimeric Anti-Cd20 Monoclonal Antibody Therapy For Relapsed Indolent Lymphoma - Half of Patients Respond to a Four-Dose Treatment Program. J Clin Oncol. 1998;16:2825–33.

Milpied, N., Vasseur, B., Parquet, N., Garnier, J. L., Antoine, C., Quartier, P., Carret, A. S., Bouscary, D., Faye, A., Bourbigot, B., Reguerre, Y., Stoppa, A. M., Bourquard, P., de Ligny, B. H., Dubief, F., Mathieu-Boue, A., Leblond, V. Humanized anti-CD20 monoclonal antibody (Rituximab) in post transplant B-lymphoproliferative disorder: A retrospective analysis on 32 patients. Ann Oncol. 2000;11:113–6.

Neumann, F., Harmsen, S., Martin, S., Kronenwett, R., Kondakci, M., Aivado, M., Germing, U., Haas, R., Kobbe, G. Rituximab long-term maintenance therapy after autologous stem cell transplantation in patients with B-cell non-Hodgkin's lymphoma. Ann Hematol. 2006;85:530–4.

O'Brien, S. M., Kantarjian, H., Thomas, D. A., Giles, F. J., Freireich, E. J., Cortes, J., Lerner, S., Keating, M. J. Rituximab dose-escalation trial in chronic lymphocytic leukemia. J Clin Oncol. 2001;19:2165–70.

Olsen, E., Duvic, M., Frankel, A., Kim, Y., Martin, A., Vonderheid, E., Jegasothy, B., Wood, G., Gordon, M., Heald, P., Oseroff, A., Pinter-Brown, L., Bowen, G., Kuzel, T., Fivenson, D., Foss, F., Glode, M., Molina, A., Knobler, E., Stewart, S., Cooper, K., Stevens, S., Craig, F., Reuben, J., Bacha, P., et al. Pivotal phase III trial of two dose levels of denileukin diftitox for the treatment of cutaneous T-cell lymphoma. Journal of Clinical Oncology. 2001;19:376–88.

Pfreundschuh, M., Trumper, L., Osterborg, A., Pettengell, R., Trneny, M., Imrie, K., Ma, D., Gill, D., Walewski, J., Zinzani, P. L., Stahel, R., Kvaloy, S., Shpilberg, O., Jaeger, U., Hansen, M., Lehtinen, T., Lopez-Guillermo, A., Corrado, C., Scheliga, A., Milpied, N., Mendila, M., Rashford, M., Kuhnt, E., Loeffler M. CHOP-like chemotherapy plus rituximab versus CHOP-like chemotherapy alone in young patients with good-prognosis diffuse large-B-cell lymphoma: a randomised controlled trial by the MabThera International Trial (MInT) Group. Lancet Oncol. 2006;7:379–91.

Press, O. W., Eary, J. F., Appelbaum, F. R., Martin, P. J., Nelp, W. B., Glenn, S., Fisher, D. R., Porter, B., Matthews, D. C., Gooley, T., Bernstein, I. D. Phase II trial of ^{131}I-B1 (anti-CD20) antibody therapy with autologous stem cell transplantation for relapsed B cell lymphomas. Lancet. 1995;346:336–40.

Press, O. W., Eary, J. F., Gooley, T., Gopal, A. K., Liu, S., Rajendran, J. G., Maloney, D. G., Petersdorf, S., Bush, S. A., Durack, L. D., Martin, P. J., Fisher, D. R., Wood, B., Borrow, J. W., Porter, B., Smith, J. P., Matthews, D. C., Appelbaum, F. R., Bernstein, I. D. A Phase I/II trial of iodine-131-tositumomab (anti CD-20), etoposide, cyclophosphamide, and autologous stem cell transplantation for relapsed B-cell lymphomas. Blood. 2000;96:2934–42.

Press, O. W., Unger, J. M., Braziel, R. M., Maloney, D. G., Miller, T. P., Leblanc, M., Gaynor, E. R., Rivkin, S. E., Fisher, R. I. A Phase 2 trial of CHOP chemotherapy followed by tositumomab/iodine I 131 tositumomab for previously untreated follicular non-Hodgkin ymphoma: Southwest Oncology Group Protocol S9911. Blood. 2003;102:1606–12.

Rosenblum, M. Immunotoxins and toxin constructs in the treatment of leukemia and lymphoma. Adv Pharmacol. 2004;51:209–28.

Salles, G., Foussard, C., Mounier, N., Morschhauser, F., Bosly, A., Lamy, T., Haioun, C., Brice, P., Boubadallah, R., Rossi, J-F, Audhuy, B., Fermé, C., Mahe, B., Lederlin, P., Sebban, C., Colombat, P., Xerri, L. Rituximab added to CHVP+IFN improves the outcome of follicular lymphoma patients: first analysis of the GELA-GOELAMS FL-2000 randomized trial. In: Blood, editor. ASH; 2004; San Diego: Blood; 2004. p. 49a.

Smith, M. R. Rituximab (monoclonal anti-CD20 antibody): mechanisms of action and resistance. Oncogene. 2003;22:7359–68.

Traulle, C., Coiffier, B. B. Evolving role of rituximab in the treatment of patients with non-Hodgkin's lymphoma. Future Oncol. 2005;1:297–306.

van Oers, M., Klasa, R., Marcus, R. E., Wolf, M., Kimby, E., Gascoyne, R. D., Jack, A., Veer, Mvt, Vranovsky, A., Holte, H., Glabbeke, Mv., Teodorovic, I., Rozewicz, C., Hagenbeek, A. Rituximab maintenance improves clinical outcome of relapsed/resistant follicular non-Hodgkin's lymphoma in patients both with and without rituximab during induction: results of a prospective randomized phase 3 intergroup trial. Blood. 2006;108:3295–301.

Vega, M. I., Huerta-Yepaz, S., Garban, H., Jazirehi, A., Emmanouilides, C., Bonavida, B. Rituximab inhibits p38 MAPK activity in 2F7 B NHL and decreases IL-10 transcription: pivotal role of p38 MAPK in drug resistance. Oncogene. 2004;23:3530–40.

Vose, J. M., Wahl, R. L., Saleh, M., Rohatiner, A. Z., Knox, S. J., Radford, J. A., Zelenetz, A. D., Tidmarsh, G. F., Stagg, R. J., Kaminski, M..S. Multicenter phase II study of iodine-131 tositumomab for chemotherapy-relapsed/refractory low-grade and transformed low-grade B-cell non-Hodgkin's lymphomas. Journal of clinical oncology. 2000;18:1316–23.

Weng, W. K., Levy, R. Expression of complement inhibitors CD46, CD55, and CD59 on tumor cells does not predict clinical outcome after rituximab treatment in follicular non-Hodgkin's lymphoma. Blood. 2001;98:1352–7.

Weng, W. K., Levy, R. Two immunoglobulin G fragment C receptor polymorphisms independently predict response to rituximab in patients with follicular lymphoma. J Clin Oncol. 2003;21:3940–7.

Wierda, W., O'Brien, S., Wen, S., Faderl, S., Garcia-Manero, G., Thomas, D., Do, K. A., Cortes, J., Koller, C., Beran, M., Ferrajoli, A., Giles, F., Lerner, S., Albitar, M., Kantarjian, H., Keating, M. Chemoimmunotherapy with fludarabine, cyclophosphamide, and rituximab for relapsed and refractory chronic lymphocytic leukemia. J Clin Oncol. 2005;23:4070–8.

Winkler, U., Jensen, M., Manzke, O., Schulz, H., Diehl, V., Engert, A. Cytokine-release syndrome in patients with B-cell chronic lymphocytic leukemia and high lymphocyte counts after treatment with an Anti-CD20 monoclonal antibody (Rituximab, IDEC-C2B8). Blood. 1999;94:2217–24.

Witzig, T. E., Flinn, I. W., Gordon, L. I., Emmanouilides, C., Czuczman, M. S., Saleh, M. N., Cripe, L., Wiseman, G., Olejnik, T., Multani, P. S., White, C. A. Treatment with ibritumomab tiuxetan radioimmunotherapy in patients with rituximab-refractory follicular non-Hodgkin's lymphoma. J Clin Oncol. 2002;20:3262–9.

Witzig, T. E., Gordon, L. I., Cabanillas, F., Czuczman, M. S., Emmanouilides, C., Joyce, R., Pohlman, B. L., Bartlett, N. L., Wiseman, G. A., Padre, N., Grillo-Lopez, A. J., Multany, P., White, C. A. Randomized controlled trial of Yttrium-90-labeled ibritumomab tiuxetan radioimmunotherapy versus rituximab immunotherapy for patients with relapsed or refractory low-grade, follicular, or transformed B-cell non-Hodgkin's Lymphoma. Journal of clinical oncology. 2002;20:2453–63.

Witzig, T. E., White, C. A., Wiseman, G. A., Gordon, L. I., Emmanouilides, C., Raubitschek, A., Janakiraman, N., Gutheil, J., Schilder, R. J., Spies, S., Silverman, D. H., Parker, E., Grillo-Lopez, A. J. Phase I/II trial of IDEC-Y2B8 radioimmunotherapy for treatment of relapsed or refractory CD20+ B-cell non-Hodgkin's lymphoma. Journal of clinical oncology. 1999;17.

13
Molecular Network Analysis using Reverse Phase Protein Microarrays for Patient Tailored Therapy

Runa Speer[1], Julia Wulfkuhle[2], Virginia Espina[2], Robyn Aurajo[2], Kirsten H. Edmiston[3], Lance A. Liotta[2], and Emanuel F. Petricoin III,[2*]

13.1 Introduction

The practice of medicine has always aimed at individualized treatment of disease. The relationship between patient and physician has always been a personal one, and the physician's choice of treatment has been intended to be the best fit for the patient's needs. The necessary pooling/grouping of disease families and their assignment to a number of drugs or treatment methods has, consequently, led to an increase in the number of effective therapies. However, given the heterogeneity of most human diseases, and cancer specifically, it is currently impossible for the treating clinician to effectively predict a patient's response and outcome based on current technologies, much less the idiosyncratic resistances and adverse effects associated with the limited therapeutic options.

Now medical research is reemphasizing the initial goal of an individualized therapy and the first glimpses of strategies and technologies to personalize medicine, based on latest advances in molecular biology, are beginning to emerge. The deciphering of the human genome is accompanied by a concomitant growth in bioinformatics. We now have an ever-increasing tool kit that helps to sift through and sort the huge amount of data generated and produce new subtle and overt correlates with functional output of cells, tissues, and organs – in both diseased and healthy states. Together with high throughput genomic and proteomics techniques, this data and the means to analyze the data will likely pave the way to the revolution in biology and medicine that we are beginning to experience today.

---

[1] University of Tübingen, Faculty of Medicine, Department of Obstetrics and Gynecology, Calwer Str. 7, 72076 Tübingen, Germany

[2] Center for Applied Proteomics and Molecular Medicine, George Mason University, Manassas, VA

[3] Department of Surgery, Inova Fairfax Hospital Cancer Center, 3300 Gallows Road, Falls Church, VA

*Corresponding Author and current affiliation for all authors:
Emanuel F. Petricoin, George Mason University, Center for Applied Proteomics and Molecular Medicine, 10900 University Blvd. MS 4E3, Discovery Hall Room 181A, Manassas, VA 20110, Phone: 703-993-8646, Fax: 703-993-4288, Email: epetrico@gmu.edu

In a highly dynamic research landscape we are able to analyze complex biological systems and to continuously refine "omics"-techniques and methods that promise to elicit major advances in personalized medicine in the near future. Despite this promise, a major question still remains, however: when and how will these technological advances make their way to the clinic so that personalized therapy will be become a reality?

13.2 Molecular Heterogeneity and the Promise of Proteomics

Cancer is a model disease for studying this question, with a significant share of the population affected, and the challenge to treat cancer patients underpinned by the complexity and heterogeneity of the disease (Sjoblom, et al. 2006). Most cancers are driven by and dependent upon multiple genetic and genomic changes as well as by varieties of aberrant signalling pathways, because genetic alterations like amplification of oncogenes or loss of tumor suppressor genes cause malignant transformation, as well as changes and interferences in the functional executive level of a cell's proteomic equilibrium do. The heterogeneous nature of cancer helps to explain unpredictable responses to existing drug therapies observed to date. These characteristics make a malignant disease a prime target for new proteomic molecular technologies, expediting the applications of basic research findings into daily clinical practice through translational research.

The unique signature of each individual patient's malignant disease lies in the molecular level of the cancer cell's genome – which is considered a rather static entity – and its proteome – which is more complex and dynamic than the genome, which includes splice variants, post-translational modifications and cleavage products and which is characterized by a wide dynamic range of protein expression expanding over several orders of magnitude. The dynamic nature of the proteome allows us to closely monitor changes in the state of a cell, tissue or organism over time and, therewith, to follow the course of a disease and track its pathogenic mechanisms as well as its response to therapy. Most importantly, however, is that the proteome offers us the ability to look at the functional endpoints of cell growth, death, invasion and metastasis, and this functional view provides direct analysis of the drug targets.

New classes of molecular diagnostics and molecular classification techniques are building upon traditional pathological techniques for tissue analysis over the past decade. In oncology overall histopathological measurements such as tumor size, degree of differentiation, degree of metastases, and cytogenetic analysis, are being supplemented with new clinically important analytes such as HER-2/neu. However, while these newer analytes are playing an expanding role in therapeutic decision-making, they do not begin to address or solve the complexity and heterogeneity in individual tumors that can lead to success or failure of a targeted therapeutic agent. Molecular classification using gene expression microarray profiling has demonstrated considerable potential for molecular classification

(Segal and Friedman, and Kaminski and Regev and Koller 2005; Brennan 2005). However transcript profiling provides an incomplete picture of the ongoing molecular network for a number of clinically important reasons. As previously stated, while it is certain that genetic mutations cause certain human diseases such as cancer, the analysis of the genome or transcriptome (i.e., gene transcript profiling) has never been shown to be able to predict or measure the protein signalling pathway networks that ultimately produce the cancer. The idea of a gene network itself is in many ways a misnomer, as genes do not really form interacting networks; it is the protein gene product that forms the linked and interacting enzymatic networks. Indeed, gene transcript levels have not been found to significantly correlate with total protein expression for many analytes – much less with the phosphorylated or otherwise functional forms of the encoded proteins – and provide little information about protein–protein interactions and the state of the intracellular signaling pathways (Shankavaram and Reinhold and Nishizuka and Major and Morita and Chary and, Reimers and, Scherf and Kahn and Dolginow and Cossman and Kaldjian and Scudiero and Petricoin and Liotta and Lee and Weinstein, 2007). Finally, most new molecularly targeted therapeutics are directed at protein targets (i.e., HERCEPTIN and GLEVEC target c-erbB2 and c-kit/abl/PDGF proteins, respectively), and these targets in the cellular proteome are constantly fluctuating depending on the cellular micro-environment, tissue micro-ecology and patient-specific alterations.

Proteomic approaches offer powerful technologies for protein separation and identification, for characterizing their biomolecular interactions, function and regulation, as well as for storing and distributing protein information (Aebersold and Goodlett 2001; Chakravarti, Chakravarti and Moutsatsos 2002; Figeys 2002; Panisko, Conrads, Goshe and Veenstra 2002). In the context of translational research proteomics emphasis is on the analysis of clinically relevant biological samples like biopsy specimens and body fluids in order to find key nodes in a diseased tissue and approach the ultimate goal – patient tailored therapies (Wulfkuhle, Edmiston, Liotta and Petricoin 2006). Until now, a missing key analytical component for the effective molecular analysis of human tissue, and ultimately the personalization of therapy, was the ability to generate a portrait of the activity of the cellular "circuitry"– the actual drug targets that are the signalling pathways and molecular networks within a cell from tissue from clinical biopsy material. The development of certain classes of protein microarrays is looking to overcome this hurdle.

13.3 Protein Microarrays for Molecular Analysis

The effective combination of diagnostic platforms with targeted therapeutics is optimal when the diagnostic platform provides information about the drug target itself and thus identifies and stratifies a given patient population into patients who will likely respond to a given therapy. In the real world, this analysis would not come from analysis of cell culture or animal models, but through the direct analysis of target cells taken from the patient through biopsy procurement. Most proteomic technologies have significant technological limitations due to overall analytical sensitivity, and these limitations are dramatically magnified when the analysis is of the very small biopsy samples. Most tissue biopsy specimens contain only a few thousand cells. Widely used proteomic platforms such as two-dimensional polyacrylmide gel electrophoresis, isotope-coded affinity tagging (ICAT) or differential tagging coupled with multidimensional LC-MS platforms, and antibody arrays, require relatively large numbers of cells for any significant results–many orders of magnitude greater than the quantity procured during a clinical biopsy. Currently, there is no direct PCR-like technology for amplifying proteins, thus new technologies are needed that can utilize microscopic amounts of cellular material.

In their basic form, protein microarrays are simply proteins immobilized within distinct regions, or predefined zones on a substratum with a solid support (Liotta, Espina, Mehta, Calvert, Rosenblatt, Geho, Munson, Young, Wulfkuhle and Petricoin 2003; Haab, Dunham and Brown 2001; Macbeath 2002; Zhu and Snyder 2003; Wilson and Nock 2003; Humphery-Smith, Wischerhoff and Hashimoto 2002). The immobilized material may be heterogeneous or homogeneous in nature; it may consist of cell or phage lysates, body fluids, an antibody, body fluid, or recombinant/expressed proteins. The immobilized molecules are then queried by probing with an analyte-specific molecule (i.e., antibody) that is coupled to a second signal-generating molecule such as a tagged antibody, ligand, serum or cell lysate. The signal can be from any number of chemiluminescent, colorimetric, fluorescent, radiometric or electrochemical read-outs. The resulting signal intensity of each spot is, in theory, proportional to the quantity of applied tagged molecules bound to the bait molecule. The spot pattern image is then captured, analyzed and correlated with biological information.

The reverse phase protein microarray (RPMA) format (Paweletz, Charboneau, Roth, Bichsel, Simone, Chen, Han, Gillespie, Emmert-Buck, Petricoin and Liotta 2001) is a defined class of protein arrays, and unlike forward phase arrays (i.e., antibody array), the RPMA is comprised of an immobilized cellular lysate, or body fluid that is directly spotted on the array surface, and then analytes are quantified by exposing the array with a specific primary antibody (Fig. 13-1). Signal amplification is independent of the immobilized protein, permitting the coupling of detection strategies with highly sensitive tyramide amplification chemistries (Bobrow, Shaughnessy and Litt 1991). RPMA technology is well suited and, in fact, specifically developed for clinical proteomics and clinical applications. Currently, one of the most important attributes of the platform lies in the ability to provide a "map" of the state of multiple in vivo cell signalling pathways, and to provide critical information about protein post-translational modifications, such as the phosphorylation states of these key proteins (Wulfkuhle, et al. 2005; Paweletz, et al. 2001; Liotta, et al. 2003; Petricoin, Wulfkuhle, Espina and Liotta 2004). The post-translational modifications reflect the functional activity state of signal pathways and networks; phosphorylation-driven information that cannot be generated from gene transcript profiling or DNA analysis. Identification of disease-related

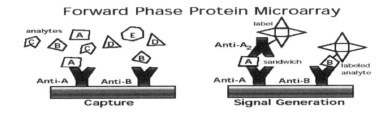

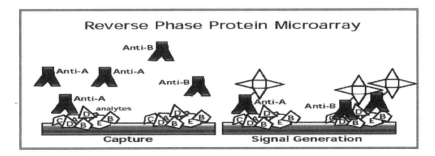

Fig. 13-1 Typical format for forward (top) and reverse (bottom) phase protein microarrays. The reverse phase array is constructed by printing material comprised of the analyte itself, usually in the form of a whole cell lysate or body fluid. An important attribute of the reverse phase array is the requirement for only one primary antibody, and the use of third generation immunoassay amplification chemistries such as tyramide precipitation. This combination greatly expands the number of analytes that can be measured in a given sample

alterations with the kinase-driven cellular networks is a key starting point for both drug development as well as design of individual therapy regimens since so many targeted therapies are kinase inhibitors and protein-protein interaction disruptors; analysis of cell signalling and signalling networks, unlike genomic endpoints, represent the drug targets themselves. RPMA may be used to monitor changes in protein phosphorylation over time, before and after treatment, between disease and non-disease states and responders versus non-responders, allowing one to infer the activity levels of the proteins in a particular pathway in real time to tailor treatment to each patient's cellular 'circuitry."

13.4 Design of Reverse Phase Protein Microarrays for Clinical Use

Because human tissues are composed of hundreds of different interacting cell populations, they provide both a unique opportunity as well as a significant barrier in the quest for discovery of disease-related information that reflects the cellular micro-environment. However, technologies such as Laser Capture Microdissection (LCM) now provide a means for routine and facile procurement of desired cell populations directly from tissue sections, while maintaining the overall fidelity of the cellular DNA, RNA and protein molecular content (Emmert-Buck, et al. 1996). Combining LCM with RPMA technology provides a fairly rapid approach for isolating pure cell populations from heterogeneous human biopsy specimens, generating a protein lysate, and spotting this lysate onto nitrocellulose-coated slides using a robotic arrayer. The resulting RPMA can thus be composed of hundreds of patient samples (or cellular lysates from cell lines). Each array is then incubated with a single primary antibody (i.e., anti- phospho-ERK kinase) and a single analyte endpoint is measured and directly compared across multiple samples on the same slide (Fig. 13-2). During printing, each patient sample is arrayed as a series of dilutions – usually at 1:2 – providing an internal standard curve. This printing strategy is important in providing a direct quantitative measurement once the linear range of detection is established. Positive and negative controls, along with calibrators, are also printed concurrently, providing a straightforward quantitative analysis and inter/intra assay comparisons (Fig. 13-2). In addition, the RPMA is a flexible insofar as non-denatured lysates can also be directly printed, so that protein–protein, protein–DNA and/or protein–RNA complexes can be detected and characterized.

Key technological components of the RPMA offer unique advantages over other array based platforms such as tissue arrays (Giltrane and Rimm 2005) or antibody

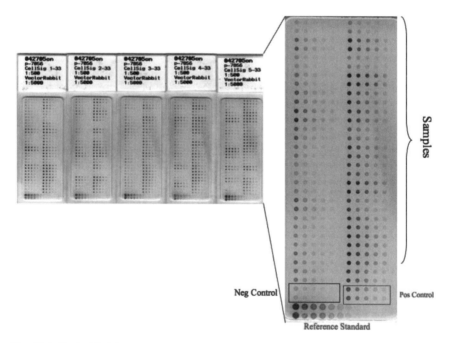

Fig. 13-2 Typical RPMA. A series of experimental samples are arrayed in a series of duplicate dilution curves (here as a series of five 1:2 dilutions, from left to right). On every slide a positive and negative control is printed along with a reference standard calibrator. These standards serve to unitize the intensity values and normalize results across slides and experiments

(forward phase) arrays. The RPMA can and does routinely employ denatured lysates so that antigen retrieval, a significant limitation for tissue arrays (especially using phospho-specific antibodies), antibody arrays, and immunohistochemistry technologies, is not an issue. Since RPMAs require a single class of antibody per analyte protein and do not require sandwich assays or direct tagging of the protein as readout for the assay, the format can use the thousands of specific antibodies such that a huge number of analytes can be measured. Other technologies, such as suspension bead array platforms, have significant limitations in the portfolio of analytes that can be measured, even in multiplex, because of the requirement of a two-site assay. Most importantly, however, is that the RPMA format can provide a multiplexed portrait of analyte levels from a very small number of cells, with over 200 endpoints being measured from 10,000 to 20,000 cells (Liotta, et al. 2003). The ability to generate quantitative data from minute quantities of cellular input without a two-site assay also enables a marked improvement in reproducibility, sensitivity and robustness of the assay over other techniques (Liotta, et al. 2003).

13.5 Signal Pathway Profiling of Human Cancer Using RPMA

Recent case studies demonstrate the utility of RPMA for the analysis of surgically obtained tissues and thereby demonstrate the potential of this format for aiding in therapeutic decision-making by providing information about the activity of signalling proteins. The first published demonstration of RPMA analyzed human prostate surgical specimens, and revealed that members of the PI3 kinase/pro-survival protein pathways are activated at the invasion front during prostate cancer progression (Paweletz, et al. 2001). In another study, the investigators examined the differences in pro-survival signalling between Bcl-$2^{+/-}$ lymphomas (Zha, et al. 2004). Comparison of various pro-survival proteins in Bcl-2^+ and Bcl-2^- follicular lymphoma subtypes by reverse phase protein microarrays suggested that there are pro-survival signals independent of Bcl-2. RPMA analysis has also been applied to evaluation of cellular signalling within the stromal and epithelial compartments from patient-matched prostate cancer study sets (Grubb, et al. 2003). Ovarian and breast cancers have also been analyzed for cell signalling activation fingerprints using this approach (Wulfkuhle, et al. 2003; Petricoin, et al. 2005). These studies reveal the patient-specific nature of signalling at a molecular level: each cancer presents a unique constellation of cellular signalling. However, that being said, signalling class-specific groupings (i.e, mTOR pathway activation) are also observed (Petricoin, et al. 2005) and indicate that a new type of molecular classification of human cancer built on functional signalling activity may be possible.

Reverse phase protein microarrays have also been used to compare cell-signalling portraits in patient-matched primary and metastatic cancer lesions (Petricoin, et al. 2005; Sheeehan, et al. 2005). Because the tissue micro-ecology of the metastatic lesion is inherently different from the environment within the primary tumor, cell signalling events may be significantly altered depending on the site of metastasis.

Since the signalling changes in the metastasis would be the most appropriate for the selection of targeted therapy due to the fact that metastasis most often determines mortality, it might be critical to develop a profile of metastatic cells themselves. This appears to be so, as these initial studies indicate a significant difference between the patient-matched primary and metastatic lesion. In the future, a patient that presents with multiple metastatic sites could be treated with a selected combination of different targeted therapies, tailored to the different signalling changes.

13.6 A Vision for Patient Tailored Therapy Using Proteomics

Molecular profiling of the human "kinome", and ongoing signalling cascades holds great promise in effective selection of therapeutic targets as well as selection and enrichment of therapy-responsive patient (Fig. 13-3). Armed with a greater appreciation of the individuality of human disease process at the molecular level, we may be able to overcome the frustratingly unpredictable nature of response to cancer to current therapies. Phosphorylation-driven signal transduction pathway profiling is particularly useful in this area since these endpoints are the direct drug targets themselves. The generation of a portrait of the state of these networks will provide the data necessary for a rationally based formulation of targeted therapies, perhaps in combination with each other. Monitoring different phosphoprotein levels will also help to identify treatment-acquired resistance to chemotherapy. Perhaps the

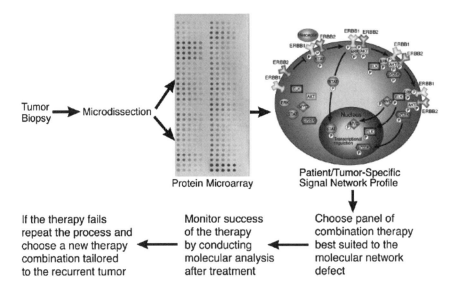

Fig. 13-3 Individualized Cancer Therapy. Following tissue acquisition by biopsy or needle aspiration, RPMA phosphoproteomic analysis is performed and signalling profile becomes the basis of a patient-tailored therapeutic regime. Therapeutic efficacy can be assessed by re-biopsy and signalling analysis

greatest promise of this technology is the opportunity to identify the earliest disease pathway changes and provide information that could lead to prevention. For example, molecular network analysis of cancer stem cells or pre-malignant lesions could provide adistinct signalling portrait that could truly cure the disease. The promise of proteomic-based profiling and, distinct from gene transcript profiling, is that the underpinning signatures are derived from the functional drug targets (i.e., activated kinases), not genes, so the pathway analysis provides a direction to therapy. In effect, the phosphoproteomic pathway analysis becomes both a diagnostic/prognostic signature as well as a guide to therapeutic intervention.

However, before the widespread adoption of phosphoproteomic-based molecular endpoints for routine clinical practice, several challenges will need to be overcome. Focused and standardized specimen acquisition is critical to ensure that low abundance proteins and post-translationally modified isoforms are kept intact and can be correctly identified. Optimally, measurements will have to be validated in large clinical trials and, at the least, will have to be performed in a CAP/CLIA laboratory setting, something that is now occurring. Under CAP/CLIA regulations, the RPMA will require the development of reference standards, controls and calibrators, and measures of proficiency. Changes in tissue fixation processes and pathological workflow will also have to occur for routine phosphoprotein-based analysis. Currently, formaldehyde fixation and paraffin embedding is the standard method for tissue preservation. This procedure is at odds with molecular analysis using RNA and protein endpoints since fixation takes place over days, not minutes. During this fixation time, gene and protein expression levels change dramatically as the living tissue is slowly fixed. Current proteomic analysis requires snap freezing of tissues to avoid cross linking of proteins and the resulting difficulties of extraction from fixed material. The requirement of frozen material for molecular analysis can be exceedingly challenging from a logistical perspective in clinical offices and practices where blood is drawn and biopsies are performed. Many clinical practices do not have infrastructure for immediate freezing. However, new classes of embedding material and the growing use of alcohol fixation as a replacement to formalin (precipitation methods such as alcohol fixation do not adversely effect final protein yield) illustrate changes that are taking place in routine pathology departments that can accelerate the use of molecular information for clinical decision-making. The confluence of new molecular discoveries using RPMA and other types of molecular profiling techniques, along with changes in pathology practices where pathologists become molecular diagnosticians, are producing a paradigm shift towards the personalization of medicine.

References

Aebersold, R., Goodlett, D. R. (2001) Mass spectrometry in proteomics. Chem Rev 101(2), 269–95.
Bobrow, M. N., Shaughnessy, K. J., and Litt, G. J. (1991) Catalyzed reporter deposition, a novel method of signal amplification. II. Application to membrane immunoassays. J Immunol Methods. 137(1), 103–12.

Brennan, D. J. et al. (2005) Application of DNA microarray technology in determining breast cancer prognosis and therapeutic response. Expert Opin Biol Ther 5, 1069–1083.

Chakravarti, D. N., Chakravarti, B., Moutsatsos, I. (2002) Informatic tools for proteome profiling. Biotechniques Suppl 4–10: 12–5.

Emmert-Buck, M. R., et al. (1996) Laser capture microdissection. Science 274, 998–1001 (1996).

Giltrane, J. M. & Rimm, D. L. (2005) Technology insight: identification of biomarkers with tissue microarray technology. Nature Clin Pract Oncol 1, 104–111.

Grubb, R. L., et al. (2003) Signal pathway profiling of prostate cancer using reverse phase protein microarrays. Proteomics 3, 2142–2146.

Haab, B. B., Dunham, M. J., and Brown, P. O. (2001) Protein microarrays for highly parallel detection and quantitation of specific proteins and antibodies in complex solutions. Genome Biol 2(2).

Humphery-Smith, I., Wischerhoff, E., and Hashimoto, R. (2002) Protein arrays for assessment of target selectivity. Drug Discovery World 4(1), 17–27.

Liotta, L. A., Espina, V., Mehta, A. I., Calvert, V., Rosenblatt, K., Geho, D., Munson, P. J., Young, L., Wulfkuhle, J., and Petricoin,, E. F. (2003) Protein microarrays: Meeting analytical challenges for clinical applications. Cancer Cell 3(4), 317–25.

Macbeath, G. (2002) Protein microarrays and proteomics. Nat Genet. 32 Suppl(526–32).

Paweletz, C. P., Charboneau, L., Roth, M. J., Bichsel, V. E., Simone, N. L., Chen, T., Han, N., Gillespie, J. W., Emmert-Buck, M., Petricoin, E. F., and Liotta, L. A. (2001) Reverse phase proteomic microarrays which capture disease progression show activation of pro-survival pathways at the cancer invasion front. Oncogene. Apr 12;20(16), 1981–9.

Panisko, E. A., Conrads, T. P., Goshe, M. B., Veenstra, T. D. (2002) The postgenomic age: characterization of proteomes. Exp Hematol 30(2), 97–107.

Petricoin, E., Wulfkuhle, J., Espina, V., and Liotta, L. A. (2004) Clinical proteomics: revolutionizing disease detection and patient-tailoring therapy. J Proteome Res. 3(2), 209–17.

Petricoin III, E. F., et al. (2005) Mapping molecular networks using proteomics: a vision for patient-tailored combination therapy. J Clin Oncol 23, 3614–3621.

Segal, E., Friedman, N., Kaminski, N., Regev, A. & Koller, D. (2005) From signatures to models: understanding cancer using microarrays. Nat Genet 37 Suppl., S38–S45.

Shankavaram, U. T., Reinhold, W. C., Nishizuka, S., Major, S., Morita, D., Chary, K. K., Reimers, M. A., Scherf, U., Kahn, A., Dolginow, D., Cossman, J., Kaldjian, E. P., Scudiero, D. A., Petricoin, E., Liotta, L., Lee, J. K., Weinstein, J. N. (2007) Transcript and protein expression profiles of the NCI-60 cancer cell panel: an integromic microarray study. Mol Cancer Ther. Mar 5.

Sheehan, K. M., et al. (2005) Use of reverse-phase protein microarrays and reference standard development for molecular network analysis of metastatic ovarian carcinoma. Mol Cell Proteomics 4, 346–355.

Sjoblom, T., Jones, S., Wood, L. D., Parsons, D. W., Lin, J., Barber, T. D., Mandelker, D., Leary, R. J., Ptak, J., Silliman, N., Szabo, S., Buckhaults, P., Farrell, C., Meeh, P., Markowitz, S. D., Willis, J., Dawson, D., Willson, J. K., Gazdar, A. F., Hartigan, J., Wu, L., Liu, C., Parmigiani, G., Park, B. H., Bachman, K. E., Papadopoulos, N., Vogelstein, B., Kinzler, K. W., Velculescu, V. E. (2006) The consensus coding sequences of human breast and colorectal cancers. Science 314(5797), 268–74.

Wilson, D. S. and Nock, S. (2003) Recent developments in protein microarray technology. Angew Chem Int Ed Engl 42(5), 494–500.

Wulfkuhle, J. D., et al. (2003) Signal pathway profiling of ovarian cancer from human tissue specimens using reverse-phase protein microarrays. Proteomics 3, 2085–2090.

Wulfkuhle, J. D., Edmiston, K. H., Liotta, L. A., Petricoin, E. F. (2006) Technology Insight: pharmacoproteomics for cancer-promises of patient-tailored medicine using protein microarrays. Nat Clin Pract Oncol. May;3(5), 256–68.

Zha, H., et al. (2004) Similarities of pro-survival signals in Bcl-2-positive and Bcl-2-negative follicular lymphomas identified by reverse phase protein microarray. Lab Invest 84, 235–244.

Zhu, H. and Snyder, M. (2003) Protein chip technology. Curr Opin Chem Biol 7(1), 55–63.

Index

A

Adenosine triphosphate (ATP), 130
Ad-TRAIL-transduced CD34+ cells, 103
AGITG. *See* Australasian Gastrointestinal Trials Group
AKT coupled assay schematic representation, 24
Alemtuzumab humanized monoclonal antibody, 167
Alpha-v/beta 8-integrin, 81
Angiogenesis
    cell-cell adhesion in, 82
    cell-matrix adhesion, 79–82
    in tumor growth, 117
Angiopoietins, 76, 78
Antibody-dependent cellular cytotoxicity (ADCC), 156
Anti-CD20-induced apoptotic signals, 156
Anti-chimeric (HACA) antibodies, 157
Anti- monoclonal antibodies (HAMA) development, 157
Anti-phosphorylated Mcm2 antibodies, 32
Antitumor-stem-cell response, 96
Aurora-A relevant kinase, 44
Aurora kinases
    Aurora-A and Aurora-B, 55–58
    functions in mitosis, 55–56
    history of, 54–55
    links with cancer, 58–59
    substrates in mitosis, 56
Australasian Gastrointestinal Trials Group, 146
A2780 xenografts, 32, 68

B

Basic Fibroblast Growth Factor, 76, 80, 82
Battista Grassi's law, 3
Becton Dickinson FACSCalibur, 45

Bioexperimental paradigm, 6, 9
Biomarker identification and validation, 31
BubR1 with anaphase promoting complex/cyclosome, 57

C

Camillo Golgi's law, 3
Cancer-associated inflammation, 112
Cancer cell-kill paradigm, 6
CAP/CLIA regulations, 185
Carcinogenesis, causes, 112, 113
Causal model, 4
CCL2 chemokine role, 112, 114, 115
CD34+ cells adenoviral transduction, 103
Cdc7 inhibition biomarkers, 31, 32
Cdc7 kinase depletion in HeLa cells effects, 22
CD20-expressing tumors, 157
CDKs inhibitors microarray analysis, 33
CD40L and IL2 vaccine, 92, 96
CD40L/IL2 B-CLL tumor vaccines, 94
CD34 TRAIL+ cells
    toxicity of, 105
    in vitro activity of, 103
    in vivo antitumor activity of, 104
CellQuest software for PF-00477736 abrogation, 45
Checkpoint kinase 1 (Chk1), 22, 43
    abrogation of, 45
    inhibitor of, 44
Chemokines in monocytes recruitment, 114–118
Chronic inflammation and cancer, 112
Chronic Lymphocytic Leukemia (CLL), 167
Claudin-5 genes, 82
Clinical-epidemiological approach, 6
Complement-dependent cytotoxicity (CDC) role, 156

CSF-1 role in mammary tumor regulation, 119
Cyclooxygenase (COX) in prostanoid biosynthetic pathway, 122
Cytokine network role in tumor site, 116

D
Darwinian models of cancer progression, 6, 10
Darwinian paradigm in oncology, 9, 10
Death receptor-dependent pathways of apoptosis, 100
DNA replicative helicase, 22
Δ PH-AKT2 assays, 24
Drug administration importance of continuing, 151
Drug discovery
  and cancer evolution, 20–21
  clinical development and regulatory affairs contribution, 35–37

E
EBV CTLs malignancy treatment in immuno-competent host, 90–92
EBV-Cytotoxic T lymphocytes (CTL), 90
EBV-specific cytotoxic T lymphocytes (CTL) administration overall tumor response rate, 97
ECOG/SWOG/CALGB study, 166
Ecteinascidia turbinata, 121
E2F transcription factor family, 33
Egfr gene mutations in EGFR, 133–136
EGFR homodimerization and heterodimerization, 129
EGFR in cancer
  ATP-binding site mutation in *egfr* gene, 133
  clinical trials and patient's reaction to drugs, 138, 139
  drug dose selection, 137, 138
  egfr abnormalities, 132, 133
  mutational location, 134
Endothelial cadherins, 82
Endothelial integrins, 79–82
EORTCISG- AGITG study, 146, 147
Epidermal growth factor (EGF), 117, 129
Epidermal growth factor receptor (EGFR)
  clinical development, 128
  inhibition strategies, 130
  usages for cancer therapeutics, 129, 130
Epstein-Barr virus, 89

Erlotinib in EGFR inhibition phase trials, 131, 132
Extra-cytoplasmatic regulated kinases (ERKs), 129

F
FCM (fludarabine/cyclophosphamide/mitoxantrone), 167
Fibronectin gene, 81
Fms/CSF1R inhibitor, 44
Follicular lymphoma
  RIT in, 163–164
  rituximab
    with chemotherapy, 160–163
    treatment, 159–160

G
Galectin-1 role in ovarian cancer, 117
Gastrointestinal Stroma Tumors (GIST), 144
  imatinib dose response in GIST in Phase I and II studies, 146
  KIT-mutations frequency in, 148
  randomized phase III studies of imatinib in, 146–147
Gefitinib in EGFR phase trials, 130–131
GELA study comparing eight cycles of R-CHOP and CHOP, 165
Gene expression microarray profiling and molecular classification, 178–179
Genetic approaches in macrophage activation, 114
Genotyping possible relevance, 147–149

H
Health conservation and cure, 2
Hematoxylin and eosin (H&E), 106
Hemi-desmosomes, 81
Hemopoietic stem cell transplantation (HSCT), 90
HER family of membrane receptors, 128–129
High dose therapy (HDT), 168–169
Histone H3 phosphorylation, 45
HIV-associated lymphoma, 168
Human carcinomas and chemokine, 115
Humoral immune system, 88
Hyaluronan receptor Lymphatic Vessel-1, 78
Hypoxiainducible factor (HIF), 120

I

IGF-1R kinase inhibitors, 29
IL-10 cytokine role, 116
Imatinib
    dose response in GIST in Phase I and II studies, 146
    side effects prognostic factors, 147
Immunoblastic lymphoma, 90
Immuno-suppressive cytokines, role, 116
Individualized cancer therapy, 184
Indoleamine 2,3-dioxygenase (IDO) in T cell inactivation, 122
Inflammatory components and cancer, 112
Insulin-like growth factor (IGF-I) type I, 29
Integrin expression in vascular endothelial cells, 80
Interleukin-8, 76
Isotope-coded affinity tagging (ICAT), 180
Italian Sarcoma Group (ISG), 146
^{131}I-Tositumomab myeloablative doses, 169

J

Junctional Adhesion Molecule-A genes, 82

K

Kinase-driven cellular networks, 181
Kinase selectivity screening (KSS), 39
Kinase targeted compound library (KTL), 39
    inhibitors, 145
    mutation and progression-free survival, 149
Kinase technology platform in NMS, 39
Ki-ras Lat2 model and magnetic resonance imaging, 34, 36
Knudson's analysis, 12
Kras mutations in EGFR, 135–137

L

Laser Capture Microdissection (LCM), 181
Linomide in tumor regulation, 120
Lymphangiogenesis, 75

M

MAb and high dose therapy (HDT) with autologous stem cell transplantation (ASCT), 168–169
Macrophage
    activation and polarization, 113, 114
    at tumor site, 114–116
Macrophage-colony stimulating factor (M-CSF), 115, 116
Mammalian target of rapamycin (mTOR), 135
Mammary tumor progression regulator, 119
Mannose receptor (MR), 117
Mantle cell lymphoma, 167
Marginal zone lymphoma, 166
Maximum tolerated dose (MTD), 137
Mcm2-7 complex, 22
c-Met
    kinase inhibitors, 43
    phosphorylation inhibition, 50
    subfamily of RTKs, 48
Metastatic disease second line treatment, 151–152
MInT study, 166
Mitogen-activated protein kinases (MAPK), 129
Monoclonal antibodies (MAbs), 130
Monoclonal antibodies mechanisms of action, 155–157
Monocytemacrophage lineages, 113
Mononuclear phagocytes, properties, 113
Mucosaassociated lymphoid tissue (MALT) lymphoma, 166
Multivariate model prognostic factors, 148
Myeloid suppressor cells (MSC), 118, 119

N

Neodarwinism, 4
Nerviano Medical Sciences (NMS), 20
NF-κB activation in TAM, 118
NMS project pipeline, 37–39
NOD/SCID mice bearing subcutaneous tumor nodules, 105
Non-Hodgkin's lymphoma, 155
Non-obese diabetic/severe combined immuno-deficient (NOD/ SCID) mice xenograft, 104
Non-small-cell lung cancer (NSCLC), 120, 130
    in EFGR, 132–137
Normal and malignant Side Population (SP) cells, 95
NPAT transcription factor, 33

O

Occludin genes, 82
Osteoprotegerin (OPG), 101

P

Patient-tailored therapeutic regime, 184
PDGF-B-deficient endothelial cells, 78
PDGFR-beta-positive pericyte progenitors, 78
PF-00477736 ATP-competitive inhibitor, 44
 CellQuest software for abrogation, 45
 optimal dosing schedule for, 47
 in vitro cytotoxicity of, 46
PHA-739358 small molecule inhibitor
 antitumor activity of, 65
 biochemical and cellular activities of, 61–65
 biomarker modulation, 66–68
 pharmacokinetics properties of, 68
 Western blots of tissue from mice treated with, 67
Phosphorylation-driven signal transduction pathway profiling, 184
PK/PD model, 68
Placenta-derived Growth Factor (PlGF), 76
Placenta-derived growth factor (PlGF), 116
Platelet-Derived Growth Factor (PDGF)-A, 76
Platelet Endothelial Cell Adhesion Molecule-1, 82
PLK-1 inhibitors, phenotypic profiling, 29
PLK-1 serine/threonine protein kinase, 24
Post-transplant lymphoproliferative disease, 167
Profiling techniques in macrophage activation, 114
Progression-free survival (PFS), 156
Protein kinase C (PKC), 129
Protein microarrays for molecular analysis, 179–181
Proteomics
 and molecular heterogeneity, 178–179
 and patient tailored therapy, 184–185
Prox1 transcription factor, 78

R

RAS-inducible melanoma nucleus, 16
Reactive oxygen intermediates (ROIs), 116
Reverse phase protein microarray (RPMA)
 clinical use, 181–183
 format, 180
 phosphoproteomic analysis, 184
 signal pathway profiling of human cancer, 183–184
RIT in follicular lymphoma, 163–164
Rituximab
 action mechanisms, 157
 in diffuse large B-cell lymphoma, 165
 follicular lymphoma, 159–163

S

S and G_2/M checkpoints, 44
Scientific medicine, epistemological evolution, 6–7
Secreted protein, acidic and rich in cysteine (SPARC), 121
Signalling cascades, 184
Signal pathway profiling of human cancer using RPMA, 183–184
siRNA-mediated down regulation, 22
Small lymphocytic lymphoma, 166
Small molecule Aurora kinase inhibitors
 AZD-1152, 60–61
 cellular effects of treatment with, 66
 MLN-8054, 62
 VX-680/MK-0457, 61–62
Stem cell transplantation (ASCT), 168
Stroma-Derived Factor-1, 76

T

TAM chemokines and cancer causes, links, 120
Target-driven drug discovery
 identification and validation, 21–22
 process, 18
T-cell therapeutics implementation in academic environment, 97
Tetramer-positive cells after CTL generation, 93
TGFβ cytokine role, 116
Therapeutic strategies in inflammatory disorders, targets, 119
Thymidine phosporylase (TP), 117
Tie-2 antagonist, 78
Timeline of cancer therapy, 8
T315I mutation, 60
TK inhibitors (TKI), 130
TRAIL. See Tumor necrosis factor-related apoptosis-inducing ligand
TRAIL-encoding adenoviruses (Ad-TRAIL), 102
TRAMP. See Transgenic mouse prostate
TRAMP model and magnetic resonance imaging, 35
Transforming growth factor α (TTFα), 129
Transgenic mouse prostate, 65
Trial design consequences, 150–151
Tumor-associated macrophage (TAM), 112
 M2 protumoral functions, 116–118
 role, 114
 targeting
  activation, 118–119
  angiogenesis, 120

effector molecules, 122
recruitment, 119–120
survival and matrix remodelling, 121
Tumor cells, targeting side population with tumor vaccines, 94–96
Tumor necrosis factor-related apoptosis-inducing ligand, 100
Tumor reactive T-cells frequency, 92
Tumors, angiogenic micro-environment, 75–76
Tumor stem cells, targeting, 92–94
Two-hit model, 12
Type IV collagen-derived tumstatin, 81
Type 3 latency tumors treatment, 90
Tyrosine kinase (TK) activity in the carboxy-terminal tail, 129

U

Unphosphorylated IGF-1R kinase domain ribbon diagram, 30
U2OS cells, endoreduplication induction, 66
urokinase-type plasminogen activator (uPA), 118

V

val57 polymorphism, 59
Vascular endothelial cells, integrin expression in, 80
Vascular Endothelial Growth Factor (VEGF), 115
pathway, 76–77
receptors and ligands of, 77
VE-cadherin-dependent regulation of vessel formation, 83
VE-cadherin gene, 83
VEGF-C produced TAMs role in human cervical cancer, 117
VEGFR2 phosphorylation in tyrosine, 83

W

Wnt family genes role in embryonic development, 119

Y

Yttrium-90, 157

Z

Zonula Occludens-1, 82

Printed In The United States Of America